3rd edition

geog.3

Great Clarendon Street, Oxford OX2 6DP

Oxford University Press is a department of the University of Oxford.
It furthers the University's objective of excellence in research,
scholarship, and education by publishing worldwide in

Oxford New York

Auckland Cape Town Dar es Salaam Hong Kong Karachi
Kuala Lumpur Madrid Melbourne Mexico City Nairobi
New Delhi Shanghai Taipei Toronto

With offices in

Argentina Austria Brazil Chile Czech Republic France Greece
Guatemala Hungary Italy Japan Poland Portugal Singapore
South Korea Switzerland Thailand Turkey Ukraine Vietnam

First published 2002
Second Edition 2005
Third Edition 2009

British Library Cataloguing in Publication Data

Data available

ISBN 978 0 19 913495 3

10 9 8 7 6 5

Printed in Singapore by KHL Printing Co Pte Ltd.

Paper used in the production of this book is a natural, recyclable product made
from wood grown in sustainable forests. The manufacturing process conforms
to the environmental regulations of the country of origin.

Acknowledgements

The publisher and authors would like to thank the following for permission to use photographs and other copyright material:

p4 Guy Mansfield/Panos; p6tl Paul Kline/Istockphoto; p6tr Hartmut Schwarzbach/Argus/Still Pictures; p6bl Anders Ryman/Alamy; p6br Maximilian Weinzierl/Alamy; p7 Jodi Baglien Sparkes/Shutterstock; p8tl Corel/Oxford University Press; p8tc RoseMarie Gallagher; p8tr Alayung Thaksin/Panos; p8cl Corel/Oxford University Press; p8c Ariadne Van Zandbergen/ Photographers Direct; p8cr RoseMarie Gallagher; p8bl RoseMarie Gallagher; p8bc LWA-Sharie Kennedy/Corbis UK Ltd.; p8br RoseMarie Gallagher; p9 Corbis UK Ltd.; p10tl Eye Ubiquitous/ Robert Harding; p10tc Natasha Owen/Fotolia; p10tr William F. Campbell/Contributor/Time & Life Pictures/Getty Images; p10cl John Isaac/Still Pictures; p10cr Michael MacIntyre/Hutchison Picture Library; p10bl Michael MacIntyre/Hutchison Picture Library; p10bc Ron Giling/Still Pictures; p10br Rebecca Blackwell/Associated Press; p12 Jorgen Schytte/Still Pictures; p13 Jorgen Schytte/Still Pictures; p15l E. Guigenam-Christian Aid/Still Pictures; p15cl B. Apicella/ Photofusion Picture library; p15cr Ed Eckstein/Corbis UK Ltd.; p15r Stephanie Colvey/IDRC CRDI; p17 Anna Tully/Panos Pictures; p22t Rohan Rogers; p22b Nick Haslam/Hutchison Picture Library; p23 Liba Taylor/Panos Pictures; p24tl Orde Eliason/Photographers Direct; p24tc Jamil Bittar/Reuters; p24tr Oxford University Press ; p24bl Sherwin Crasto/Reuters; p24bc Greenshoot Communications/Alamy; p24br Pavel Rahman/Associated Press; p26tl Caroline Penn/Corbis UK Ltd.; P26tr PA Photos/EPA/Empics; p26b Jorgen Schytte/Still Pictures; p27 Jorgen Schytte/Still Pictures; p28tl Anharris/Dreamstime; p28tr Perkus/ Shutterstock; p28tc Gautier Willaume/123RF; p28cr China Photos/Stringer/Getty Images; p28bc Kheng Guan Toh/123RF; p28bl Pedro/Fotolia; p28br Grosremy/Dreamstime; p30bl China Daily China Daily Information Corp - CDIC/Reuters ; p30br China Daily China Daily Information Corp - CDIC/Reuters; p33 Stringer Shanghai/Reuters; p34 Associated Press; p35 Tom Salyer/Photolibrary; p36t Greg Baker/Associated Press; p36b Imaginechina; p37 Jason Lee/ Reuters; p38t Elizabeth Dalziel/Associated Press; p38b Kin Cheung/Reuters; p39 Jing Aiping/Shutterstock; p40 Reinhard Krause/Reuters; p41tl Bill Perry/Fotolia; p41tc Ng Han Guan/Associated Press; p41tr Richard Jones/Rex Features; p42tr Chris Stowers/Panos Pictures; p42bl Istockphoto; p42br Reinhard Krause/Reuters; p43t Louisa Lim; p43b Rebecca Blackwell/Associated Press; p44t Stringer Shanghai/Reuters; p44b Chang W. Lee/The New York Times/Redux; p45t China Daily China Daily Information Corp - CDIC/Reuters; p45b China Daily China Daily Information Corp - CDIC/Reuters; p46 Clouston/Shutterstock; p47cr Richard Jones/Rex Features; p48tr Gary/Dreamstime; p48bl Olivier/Fotolia; p48br Aly Song/Reuters; p49tl Dreamstime; p49tr Elizabeth Dalziel/Associated Press; p49bl Karel Prinsloo/Associated Press; p49br Osamu Honda/Associated Press; p50tl Eniko Balogh/ shutterstock; p50tc Paul Sakuma/Associated Press; p50tr Jon Sullivan, PDPhoto.org; p50cl Natalia Bratslavsky/Shutterstock; p50cr Ves Herman/Reuters; p50bl John Gress/Associated Press; p50bc Mario Tama/Getty Images; p50br Richard Paul Kane/Shutterstock; p52tl Paul Tessier/Istockphoto; p52br Jeff Schmaltz, MOD IS Rapid Response Team at NASA GSFC; p55tl Canstockphoto; p55tr Jim Young/Reuters; p55b U.S. Department of Defence; p56t UCR/ California Museum of Photography; p56b Justin Sullivan/Staff/Getty Images; p58 Art resource; p59t Legend of America; p59b Allen Russell/Alamy; p60 Jim Wegryn (www.jimwegryn.com); p62l Elvis Presley Enterprises; p62r Hulton-Deutsch Collection/Corbis; p63t Jupiter Images; p63b Tim Boyle/Staff/Getty Images; p64l CLASS Group; p64r Istockphoto; p65t Randy Jeide/ Shutterstock; p65b Jupiter Images; p66cr Rikkirby/Dreamstime; p66br Oxford University Press; p67tl Shutterstock; p67tc Mike Blake/Reuters; p 67r Ian Kluft, Wiki Commons; p67bl Marcio Jose Sanchez/ Associated Press; p67bc Oksana Perkins/Shutterstock; p68bl U.S. Department of Defense; p68bc U.S.Federal Government; p68br RoseMarie Gallagher;p68bg Jason Stitt/123RF; p69 Larry Downing/Reuters; p69bg Jason Stitt/123RF; p70 Dreamstime; p72tl Imagno/Contributor/Getty Images; p72tc Royal Geographical Society; p72tr M&N/Alamy; p72bl Gina Smith/Shutterstock; p72bc Mike Hutchings/Reuters; p72br Simone van den Berg/ Shutterstock; p73tl Brooks, Robin/Bridgeman Art Culture History; p73cr Oxford University Press; p74 Oxford University Press; p76tl Don Ryan/AP Photo; p76tr Anat Givon/AP Photo; p76bl Daniel O'Leary/Panos Pictures; p76br Foybles/Alamy; p78 Corel/Oxford University Press; p79 Toby Adamson/Still Pictures; p80cr Mohammad Shahidullah/Reuters; p80br Rafiquar Rahman/Reuters; p81 RoseMarie Gallagher; p82t The Photolibrary Wales/Alalmy; p82c Emyr Rhys Williams; p82b David Gibson/Photofusion picture Library; p84tl Paul A. Saudres/Corbis UK Ltd.; p84tr Charles O'Rear/Corbis UK Ltd; p84cl Bettmann/Corbis UK Ltd.; p84cr Paul A. Saudres/Corbis UK Ltd.; p84bl Lito C. Uyan/Corbis UK Ltd.; p84br Paul A. Saudres/Corbis UK Ltd.; p85tr Donald Stampli/AP Photo; p85cr Sherwin Crasto/Reuters; p86tl Andrew Gunnartz/Panos Pictures; p86tr Ron Giling/Still Pictures; p86cl Harmut Schwarzbach/ Still Pictures; p86cr Mike Williams/Peak Pictures; p86bl David Turnley/Corbis UK Ltd; p86br Ctherine Karnow/Corbis UK Ltd; p87tr Paul A. Souders/Corbis UK Ltd; p87br David Turnley/ Corbis UK Ltd; p88 Reuters/Luis Galdamez/Corbis UK Ltd; p90main Mark Henley/Panos Pictures; p90tr Henryk T. Kaiser/Rex Features; p90ur Stephanie Maze/Corbis UK Ltd; p90lr Jeremy Horner/Hutchison Picture Library; p90br Jeremy Horner/Corbis UK Ltd; p90 bl Kennan Ward/Corbis UK Ltd; p90bc John Isaac/Still Pictures; p92l Martin Rogers/Corbis UK Ltd; p92r Corbis UK Ltd; p94tr Alberto Lowe/Reuters; p94cr Oxford University Press; p94br Fairtrade.org; p95l Enzo & Paolo Ragazzini/Corbis UK Ltd.; p95r Simon Rawles/ pa.photoshelter; p96l Mike Goldwater/Alamy; p96c Andy Sacks/Getty Images; p96r PaulPaladin/Shutterstock; p97 Mariana Bazo/Reuters; p98 Mark Henley/Panos; p100tl Oxford University Press; p100cl Dave Bartruff/Jupiter Images; p100bl x-drew/Istockphoto; p102 Wolfgang Kaehler/Alamy; p103 RoseMarie Gallagher; p105 Phyl Gallagher; p107l Michela Scibilia; p107tr Joel Nito/Image Forum; Jacques Descloitres, MODIS Rapid Response Team, NASA/GSFC; p108tl Paisajes Espanoles; p108tr Paisajes Espanoles; p108b David Cumming/ Eye Ubiquitous/Hutchison; p110l James Davis Worldwide; p110r James Davis Worldwide; p112 Petershort/Istockphoto; p113tr Peru Nature Rainforest Expedition; p113cr Rm/ Shutterstock; p113br Morgan Stetler; p114cr Mike Page.co.uk; p114bl Mike Page.co.uk; p114br Mike Page.co.uk; p116 Rich Carey/Shutterstock; p118 NASA, Goddard Space Flight Center; p119tl, p119bl NOAA (oceanexplorer.noaa.gov); p119br Oxford University Press; p120tr NOAA; p121 RoseMarie Gallagher; p122-123bg Mana Photo/Shutterstock; p122tl Scott Dickerson/Istockphoto; p122tc dan_prat/Istockphoto; p122tr Oxford University Press; p122c Marine Current Turbines TM Limited; p122cr Kevin O'Hara/Photolibrary; p122bl Dodmedia.mil; p122bc Dhoxax/Shutterstock; p122br Yaron Kaminsky/Assosiated Press; 124bl NOAA; p124br Thomas Pickard Photography/Istock photo; p125tr John Hyde/Photolibrary; p125cr Blue Planet Society; p126 Gary Austin/Greenpeace; p128l David hughes/Fotolia; p128c Jennie Woodcock Reflections Photolibrary/Corbis; p128r David Tomlinson/Lonely Planet Images; p129 Alan Spencer Photography/Photographers Direct; p130l Lu Wenzheng/ Assosiated Press; p130r Klaas Lingbeek/Istockphoto; p131 Navy.Mil; p132t Dita Alangkara/ Assosiated Press; p132bl Sean Sprague/Still Pictures; p132bc Digital Vision/Photolibrary; p132br Steve McAlister/Getty Images; p133l Photofusion/Photographers Direct; p133c Leah-Anne Thompson/Shutterstock; p133r Keeleyson/Photobucket; p134tr Eric Baccega/ Photolibrary; p135tr NASA, Goddard Space Flight Center.

Illustrations are by Barking Dog Art; Matt Buckley; Stefan Chabluk; Richard Deverell; Jeff Edwards; Roger Fereday; John Hallett; Tim Jay; Richard Morris; David Mostyn; Mike Nesbitt; Tim Oliver; Colin Salmon; Mike Saunders.

The Ordnance Survey map extract on p115 is reproduced with the permission of the Controller of Her Majesty's Stationery Office © Crown Copyright. The map on page 119 is reproduced with the permission of GEBCO (The General Bathymetric Chart of the Oceans).

The publisher and authors would like to thank all the individuals and organizations who have helped during research for this book. In particular, and in topic order:

Tamsin Maunder and other staff of WaterAid; Kate Kilpatrick, Oxfam; Alex 'Walter' Middleton; the International Coffee Organization; the Office of National Statistics; Patricia Barnett and Tourism Concern, London; John Davies of The Broads Authority; Professor Martin Jakobsson of Stockholm University and GEBCO.

Thank too to our excellent reviewers who have provided thoughtful and constructive criticism at various stages: Stephen Kaczmarczyk, Caroline Jiggins, Anna King, Phyl Gallagher, John Edwards, Katherine James, Roger Fetherston, Philip Amor, and Michael Gallagher.

We would also like to thank Janet Williamson for both general and specific contributions to the geog.123 course. Thanks too to Omar Farooque for help and support.

Information has been drawn from many sources. We would like to acknowledge in particular: the keynote speech by Professor Sir John Beddington, the government's Chief Scientific Adviser, at the GovNet Communications' Sustainable Development 09 conference, March '09; WWF's Living Planet Report 2008; an article about jeans in the Guardian of 29th May 2001.

Every effort has been made to contact copyright holders of material reproduced in this book. Any omissions will be rectified in subsequent printings if notice is given to the publisher.

Cover images: Getty (globe) and Alamy.

Contents

1 Development

The big picture

This chapter is about **development** – the process of change for the better. These are the big ideas behind the chapter:

- Development is about improving people's lives.
- It goes on all over the world, in every country, including the UK.
- Every country is at a different stage of development.
- There is a big development gap between the rich and poor countries.
- Poor countries need some help from richer ones, to close the gap.

Your goals for this chapter

By the end of this chapter you should be able to answer these questions:

- Development has many different aspects. Having enough money to live on is one. Which others can I list? (At least four.)
- Where is Ghana, and what can I say about its physical features and climate? (Give at least six facts about it.)
- What are *development indicators*, and what six examples can I give?
- How developed is Ghana, compared to other countries?
- What do these terms mean, and which countries can I give as examples? (At least two different countries for each!)

  LEDC *MEDC* *Third World* *rich north* *poor south*

- What characteristics do LEDCs tend to have in common? (At least five.)
- For what kinds of reasons do countries lag behind in development? (Give at least four kinds, with examples.)
- Why is Ghana still an LEDC? (Give at least four reasons.)
- What can poor countries do, to earn money for development?
- What can rich countries do, to help poorer countries develop?
- What are the Millennium Development Goals, and what examples can I give? (Try for two examples.)

And then …

When you finish the chapter, come back to this page and see if you have met your goals!

Did you know?

If the world were a village of 100 people …
- *the 2 richest villagers would have over half the total wealth.*

Did you know?

If the world were a village of 100 people …
- *the 50 poorest villagers, between them, would have only 1% of the total wealth.*

Did you know?

If the world were a village of 100 people …
- *20 would be Chinese*
- *16 would be Indian*
- *1 would be from the British Isles.*

What if …

- *… the UK were one of the world's poorest countries?*

Your chapter starter

Look at the photo on page 4.

Something new has arrived in this village in Ghana. What is it?

Why is everyone looking so happy?

Why didn't they have this thing before (like you do)?

Do you think there are many people who still don't have it?

1.1 Rich world, poor world

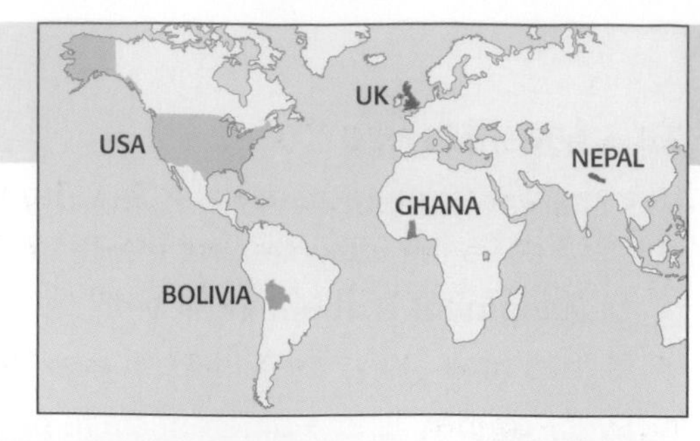

This unit is about how unequal our world is.

Comparing lives

You did not choose where you were born – but it has a huge impact on your life. Compare these four young people. They are all aged 15.

Hannah lives in the USA. She likes science, and plays the trumpet. She has just been to Mexico on a school trip. She gets $30 a week pocket money (about £20). Sometimes she wants to be an engineer, and sometimes a writer.

Joe lives in Ghana. He's top in his class at maths. He plays lots of football with his friends. He'd like to run a business, and buy a nice house for his mum. But he hopes to begin with an office job. £50 a month would be great!

Julien lives in Bolivia. He's a shoeshine boy. He earns about 6p a customer. He lives in the family shack, with no running water. But he studies every evening at a special centre. He's learning to read, and write, and use computers.

Nisha lives in Nepal. She has never been to school. She helps on the farm, and collects firewood for cooking, and looks after her brothers and sisters. She has not seen herself in a mirror for years! They had one once, but it got broken.

Not everyone in their countries lives like those four. For example in the USA, many families are much better off than Hannah's, and many much poorer.

But *overall*, people have a much higher standard of living in some countries than in others. The world is a very unequal place.

An unequal world

Of those four teenagers, Hannah has the highest standard of living.
She has water on tap, and electricity. A good education. More than enough food. Money to spend. And plenty of choices ahead of her.

So are more people's lives like hers – or like Nisha's?

Think about this:

- The world has over 6.7 billion people. (That's 6700 million.)
- About 15% of them live in abject poverty, on less than $1 (about 70 p) a day. For food, shelter, clothing, fuel, medicine, everything.
- 77% live on less than $10 a day.
- 24% do not have electricity.
- 16% do not have access to clean safe water.
- 39% do not have access to adequate toilets.
- 8% never get enough to eat, ever.
- Around 1 in 5 people aged 15 and over can't read or write.
- Around 25 000 young children die *every day*, mainly from causes linked to poverty.

So that makes Hannah a very lucky person. What about you?

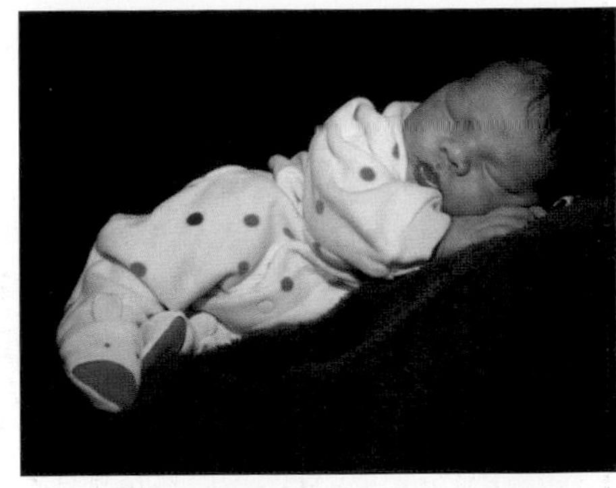

▲ *Some mums hope their babies will grow up bright and beautiful. Some just hope they'll survive.*

Did you know?

◆ *If the world were a village of 100 people: 27 of them (or more than a quarter) would be under 15 years old.*

It's all about development

In a **developed** country, almost everyone has enough food, and enough money to survive on, plus electricity, a clean water supply, and access to education, health care, and so on.

The world has over 190 countries.
Each is at a different stage of development.
The USA, and the UK, are more developed than most.

Some countries have fallen far behind, in development.
This is one of the biggest challenges facing the world today.

So in this chapter we focus on development. We'll start in the next unit by looking more closely at what development means.

Your turn

1 a Look at the four teenagers. See if you can put them in order of their standard of living, with the highest standard of living first.
Give reasons for the order you chose.
b Which one do you think has a life most like yours?

2 Look again at the four teenagers.
a Can you say which one is the happiest? Explain.
b Is it possible that Nisha could be the happiest? Give reasons for your answer.

3 Look at the statistics in the bullet points above.
a Which did you find the most surprising?
b Which do you find the most shocking?

4 Now imagine the world is a village of 100 people. From the statistics given above, how many would:
a always be hungry?
b have only dirty water to drink?
c have less than $1 a day to live on?

5 The statistics show that the world is an unequal place. Do you think it's anyone's fault? Explain.

6 Most of us are okay, here in the UK. So does it matter that the world is an unequal place? Give your reasons.

7 What kind of questions would you ask about a country, to see how developed it is? This unit will give you some ideas. But see if you can come up with others too.

So what is development ?

In this unit you'll learn what 'development' means.

It is many different things

Development is about **improving people's lives**. So it's not just about getting richer, or owning more things. It has many different aspects.

Everybody's doing it

As you saw in the last unit, the world has over 190 countries. All are striving to develop.

But some are developing very slowly – or even going backwards.

So now there's a big gap in development between the most and least developed countries. Closing the gap is a huge challenge.

Your turn

1 On page 8, the nine speech bubbles with *red* outlines show nine key aspects of development.

a Write down this heading:
Development – change for the better

b Under your heading, list the other eight key aspects of development. Put them in what *you* think is their order of importance, most important first.
(For example would you put *the chance of a good education* first ?)

c Do you think everyone in the world would choose the same order as you ? Explain your answer.
(Compare lists with a partner, to check !)

2 Look at the drawing above.

a What do you think it represents?

b What does it tell you about the UK ?

3 Development costs money. For example it costs a lot to provide a clean safe water supply for everyone.
From page 8, write down:

a four other changes you think would cost a lot

b two that may need people to change their attitudes

c two that may need a government to pass new laws.

4 The photo below was taken in Iraq in 2003, after it had been invaded, mainly by the USA and UK. War can halt a country's development, or even reverse it.
Explain why. Show your answer as a spider map.

5 Which aspects of development do you think the UK needs to do more work on ? Write a letter to the Prime Minister giving your list, and your reasons.

1.3 Now … meet Ghana

This unit introduces Ghana, the African country we explore in this chapter, to see how developed it is.

Welcome to Ghana

Welcome to Ghana, linked to the UK by history. Where you'll find …

▲ *… a warm welcome for visitors …*

▲ *… palm-fringed beaches …*

▲ *… outdoor markets …*

▲ *… some great wealth …*

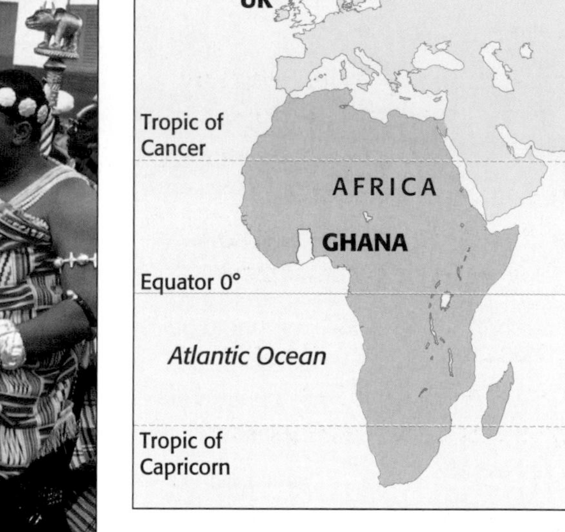

▲ *… traditional ceremonies and rituals …*

▲ *…a great sense of style …*

▲ *… much poverty …*

▲ *… and a passion for football.*

Ghana's physical geography

Much of Ghana is low and flat, as map **A** shows.

Lake Volta is an artificial lake, created when a dam was built on the River Volta, for electricity.

Accra is Ghana's capital city.

Now look at the shape of Ghana. Neat and tidy! It was carved out of separate kingdoms by the British. You can find out more about this later.

Its climate

Ghana is in the tropics, so it's warm. The south west is also wet. (It once had lots of tropical rainforest, but most has been cut down.)

The south east corner is quite dry. And as you go north, Ghana gets hotter and drier. The far north is very hot and dry, due to winds from the Sahara. It often suffers drought. Map **B** gives a summary.

A

B **Ghana's climate**

Its people and their lives

- Ghana has a population of about 23 million.
- Over half live in rural areas. But as in most countries, more and more people are moving to the towns and cities – usually to find work.
- 56% of Ghana's workforce depends on farming for a living. Thousands of farmers in the south west grow cocoa, for your chocolate!
- Ghana is quite rich in natural resources. Look at the map, and box **C**. But in spite of this, it's still quite poor. It is about the 45th poorest country in the world.
- Half the people have no electricity yet. A quarter have no access to clean safe water. They get their water from rivers and ponds.
- Thanks to poverty, people in Ghana can expect to live to be 59, on average. (For the UK, it's 79.)

But Ghana could soon be much better off. Large deposits of oil and gas were found off its coast, in 2007. These could earn it a great deal of money.

C **Its main natural resources**

- gold and diamonds
- bauxite (aluminium ore)
- oil and gas (found in 2007)
- forests (being cut down for timber)
- fish from the sea and Lake Volta
- the climate in the south west suits crops like cocoa and palm oil

Your turn

1 Where is Ghana? Use these terms in your answer: ocean, West Africa, tropic, equator.

2 Name the countries that border Ghana. (Page 141.)

3 Using only map **A**, write a paragraph about Ghana's physical features. (For example where is the highest land? How high? What about lakes? Rivers? Coast?)

4 **a** Would you say Ghana is a *developed* country? Give your reasons. (Glossary?)

b Which photos (if any) on page 10 support your answer in **a**? Give their number(s).

5 Now, using the information in this unit, and your own knowledge, and the table below, write a short piece comparing Ghana and the UK. See if you can give the *population density* for each country in your answer. (Use suitable headings, and write at least 60 words!)

Some statistics	Ghana	UK
Area (thousands of sq km)	240	245
Population (millions)	23	61
% of workforce in farming	55	under 2
Life expectancy (years)	59	79

Poverty in a Ghanaian village

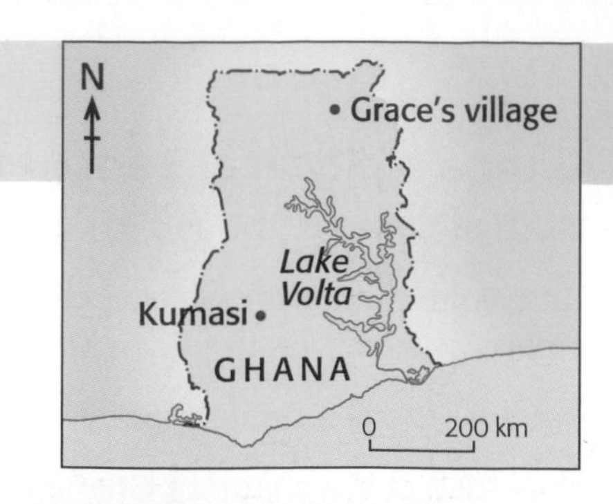

In most of the world's poorer countries, poverty is worst in rural areas. Here you'll read about poverty in a rural village in northern Ghana.

A day in the life of Grace

So you want to know what it's like to be poor ?

I lie here on my straw mat, staring up into the darkness. My baby lies beside me, snuffling in her sleep. And over there on the mud floor my four other living children, all curled up together. Out in the yard, in their graves, the two dead ones. My firstborn died when she was three, and the youngest boy last year. How I cried when I buried them.

I lie here thinking about my problems. First, my husband. A good man. He works hard, and is always thinking of ways to make our lives better. Two months ago he went to Kumasi to find work. 'We will buy a goat with the money' he said, 'and clothes for the eldest boy to go to school.' But I have had no news of him. Maybe he is ill, or in trouble.

And the farm. The rains were poor last season. Out in our tiny field the millet is dry and stunted. Enough to feed us for two months, perhaps. What then ? In the darkness I can feel my savings, tied in the corner of my cloth. Two cedis. If any of the children fall ill, that won't even be enough for medicine.

I could sell something – but what ? You could count our possessions in seconds. Three enamel bowls. Two metal plates. The cooking pot. The water bucket. The kerosene lamp made from a bottle. The wooden pestle for pounding the millet. One machete. One hoe. Two small knives. A fork. A torch with no bulb. Two mats. And a few bundles of worn clothing.

But today is a new day. Soon I will rise and slip out to the clump of bushes behind the huts, which is the village toilet. Like the other women I go while it is still dark, for privacy. And at daybreak I will set off to get water. The river is nearly dry now, so the water will be very muddy and dangerous. It killed my children. But what can I do ?

It takes me over an hour to get to the river, and longer to get back with my heavy bucket. I will give the children a little water to drink. I will breastfeed the baby. Then I will go to the farm to tend the millet and pick what's ready. And all day long I will hope that someone from the village will come running with a message from my husband.

While I am away my eldest daughter will pound millet. The eldest boy will go looking for firewood – every day a little further. Towards dusk we will eat our one meal for the day: millet porridge. At 6 it will get dark, as usual. I want to save the little kerosene that's left. So we will go to bed early, as usual – and, as usual, still hungry.

So, this is poverty. Coping with it takes all my energy. But we will survive, and I will find a way to create a better future for my children.

▲ *Grace with two of her children.*

Did you know ?
- A child dies every 10 seconds, somewhere in the world, from a disease carried by dirty water.

Did you know ?
- Ghana's currency is the cedi.
- 2 cedis = £1.08, in 2008.

▲ *Grace's village. All her friends are poor, like her. They all work very hard.*

Your turn

1 a List the items Grace has, for her kitchen.

b Now list the things in your kitchen.

2

Time spent on tasks in Grace's household

Task	*Minutes*
A preparing dinner (pounding and boiling millet, making a groundnut sauce)	200
B getting water (from the river)	170
C sweeping (the yard and hut)	45
D washing clothes (at the river)	200
E washing up (one meal a day)	20
F obtaining fuel (firewood)	120

a Make a table like this for these tasks in *your* household. (Change what's in the brackets.)

b Now draw a suitable graph to compare the times for these tasks in your household and Grace's.

c Did you have any problems in drawing the graph for **b**? If yes, explain why.

d For which task is the time difference greatest? Why?

e For which is it least? Why?

f In total, how much longer is spent on these six tasks in Grace's household than in yours? How might this affect Grace and her family?

3 Grace lives in great poverty. Draw a spider map to show what that means, for her. You could start like this.

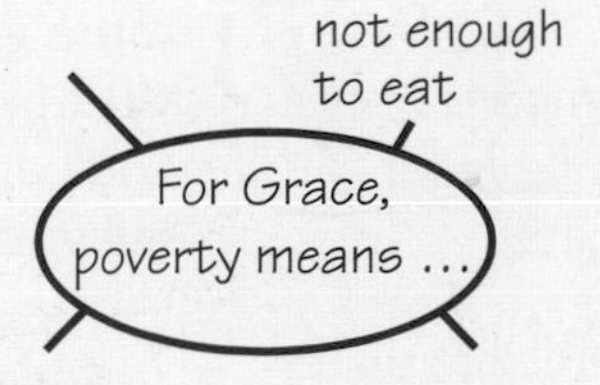

4 a Next, study the photo above, and note as many points as you can about life in Grace's village. For example what are the people doing, and using? Which groups of people are missing? Don't forget to look in the background too. What are the houses like? Are there any electricity cables?

b Now use your notes to write a couple of paragraphs about life in the village. Make them interesting!

5 Your class wants to help the people of Grace's village. You could raise money to help them to:

A install a village pump, giving clean safe water

B read and write (so Grace can write to her husband)

C fit solar cells (PV cells) to the hut roofs, so they can have electric lighting

D build a latrine (a concrete toilet where the waste drains away into the ground)

a Which do you think Grace would like first? Why? Write down all the benefits it would bring.

b Imagine you are Grace. Arrange the four projects in order of priority, from her point of view.

c Who should have most say in deciding about the projects: your class, or the villagers? Why?

6 And now it's time to tell Grace about you.

a Write a page about a day in your life, and the kinds of tasks you have to do, and what you worry about. (Imagine that someone who is able to read will read it out to Grace and her children for you.)

b How do you think Grace and her children will feel, about your life?

1.5 How developed is Ghana?

Here you'll see how data is used to measure development, and compare countries.

Measuring development

On page 8 you saw that development has many different aspects.

You've seen that Ghana has much poverty. But to get an idea of how *developed* it is, you need to ask questions like those on the right.

And then collect data to answer them!

In fact this data is collected every year, for Ghana and most other countries. It is published in tables of **development indicators**.

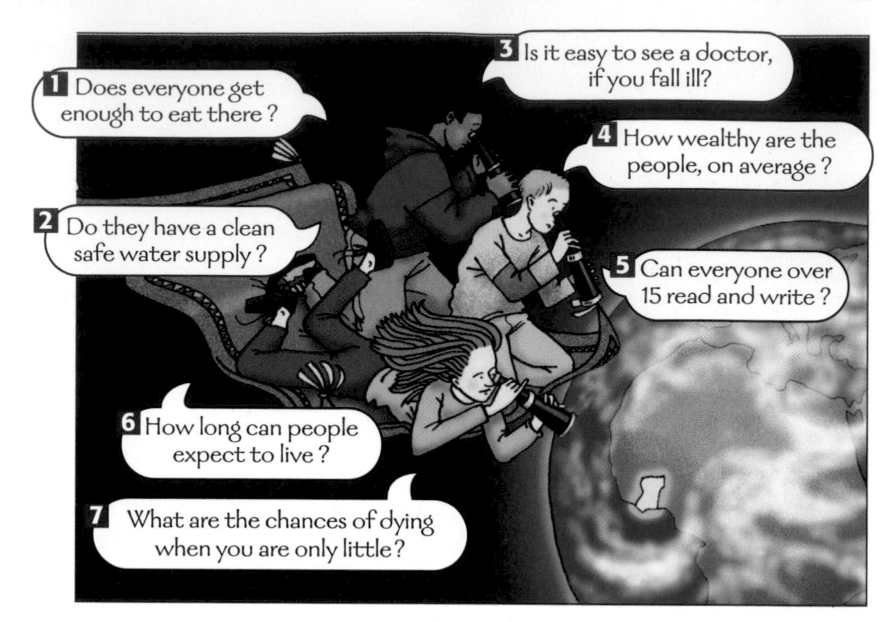

What is a development indicator?

A **development indicator** is just data that helps to show how developed a country is.

Look at question 6 above. **Life expectancy** is how long people can expect to live, on average. It is one example of a development indicator.

For people born in Ghana in 2005, life expectancy was 59 years. For people born in the UK, it was 79. So those born in the UK are likely to live 20 years longer. (You'll try to explain why, later.)

Did you know?

In every country in the world, women are likely to:
- live longer than men
- earn less than men.

Wealth as a development indicator

One indicator that's often used is **gross domestic product** or **GDP**. It's the total value of the goods and services a country produces in a year. You can think of it as the wealth the country produces.

GDP is given in **US dollars (PPP)**. (**PPP** or *purchasing power parity* means the GDP is adjusted to take into account that a dollar buys more in some countries than others.)

Dividing GDP by the population gives **GDP per capita**. This gives you a fairer way to compare countries (since some have far more people than others).

Little by little!

GDP per capita for Ghana

Year	GDP per capita (US dollars PPP)
1960	$1043
1994	$1960
2000	$1964
2005	$2480

As a country develops, it produces more goods and services. So its GDP per capita rises. Look at this table. What does it tell you about Ghana?

But GDP per capita does not tell us whether people have clean safe water to drink, for example, or enough doctors. So we need other indicators too.

Your turn

The question	The matching development indicator	Its value for Ghana in 2005
	GDP per capita	$2480 (PPP)
	life expectancy	59 years
	adult literacy rate	58%
	under-5 mortality rate	1 in 10 (or 10%)
	% with access to clean safe water	75%
	number of doctors per 100 000 people	15
	% undernourished	11%

1 The table above shows some development indicators.
   a Make a larger copy of it. (Make the first column *wide*.)
   b Now write questions 1–7, from the top of page 14, in the correct rows in the first column. (Glossary?)

2 Life expectancy is lower in Ghana than in the UK.
   a See if you can think up some reasons for this.
   b Do you think it will change as Ghana's GDP per capita changes? How? Why?

3 Next you'll compare Ghana with three other countries.
   a First make a table like this one.

	Score for ...			
	Ghana	UK	Brazil	India
Life expectancy	1	4		
Under-5 mortality rate				2
Enrolment in primary school				
Access to safe water				
GDP per capita				
Total score				

   b Now look at the data for the four baby girls below. Using this data, give each country a score of 1 – 4 for each indicator. (This has been started for you.) The country with the *best* result each time scores 4. The country with the *worst* scores 1.
   c Find the total score for each country.
   d Using the totals to help you, list the four countries in order of development, the most developed first.

Human development index (HDI), 2006

Brazil	0.807	Nigeria	0.499
Canada	0.967	Poland	0.875
China	0.762	South Africa	0.670
Ghana	0.533	UK	0.942
India	0.609	USA	0.950
Kuwait	0.912	Sierra Leone	0.329
Mali	0.391	Zambia	0.453

4 The **human development index** or **HDI** is used a lot, to indicate development. It combines data on GDP per capita, life expectancy, adult literacy, and enrolment in education, to give a score between 0 and 1. The closer to 1, the better!
   a Look at the table above. Which of those countries is best for human development? Which is worst?
   b Make a large copy of the scale on the right. Use a full page. (Graph paper?)

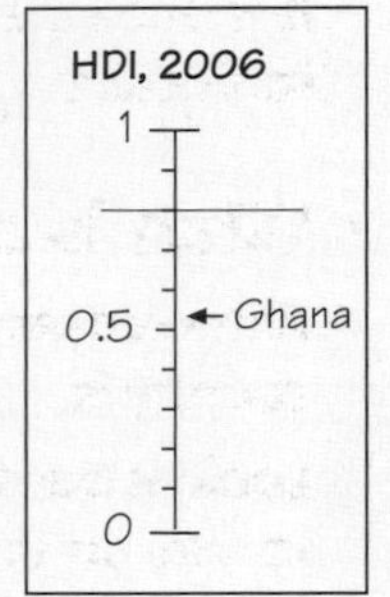

   c Mark in each country from the table, on your scale. One is in already.
   d Now draw two horizontal lines, cutting the scale at 0.8 and 0.5, as started here.
   e Above 0.8 = high human development, from 0.5 to 0.8 = medium, below 0.5 = low.
      i Shade each group of countries (high, medium and low HDI) on your scale. Use a different colour for each group, and add a colour key.
      ii To which group does Ghana belong?
      iii To which group does the UK belong?

5 So – how developed is Ghana, compared with other countries? And is it growing more developed, or going backwards? Give evidence to support your answers. (This little table may help.)

HDI for Ghana

Year	HDI
2000	0.497
2003	0.499
2005	0.524
2006	0.533

Akosua, Ghana
Life expectancy: 59
Her chances of –
dying before age 5: 11.2%
going to primary school: 65%
a safe water supply: 75%
GDP per capita: $2480 (PPP)

Molly, UK
Life expectancy: 81
Her chances of –
dying before age 5: 0.6%
going to primary school: 100%
a safe water supply: 100%
GDP per capita: $33 240 (PPP)

Maria Teresa, Brazil
Life expectancy: 76
Her chances of –
dying before age 5: 3.3%
going to primary school: 95%
a safe water supply: 90%
GDP per capita: $8400 (PPP)

Priya, India
Life expectancy: 65
Her chances of –
dying before age 5: 7.4%
going to primary school: 85%
a safe water supply: 86%
GDP per capita: $3450 (PPP)

1.6 Mapping development around the world

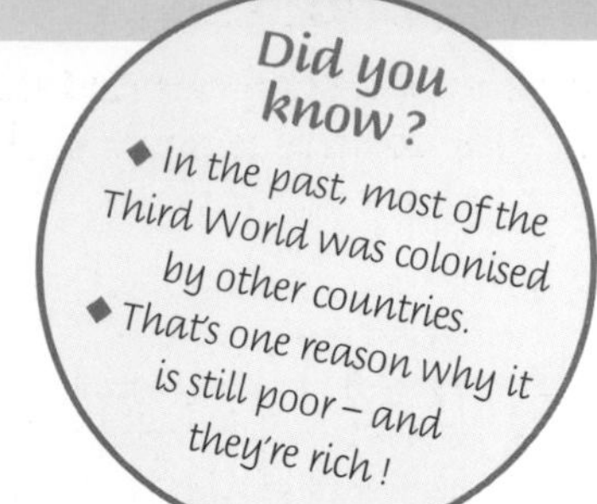

Here you'll see how an indicator can be mapped, to compare development around the world. And then you'll take a look at less developed countries.

An unequal world

As you saw in Unit 1.1, the world is a very unequal place.
You can show just how unequal it is by mapping a development indicator.
Look at this map. It shows how **GDP per capita** (PPP) varies.

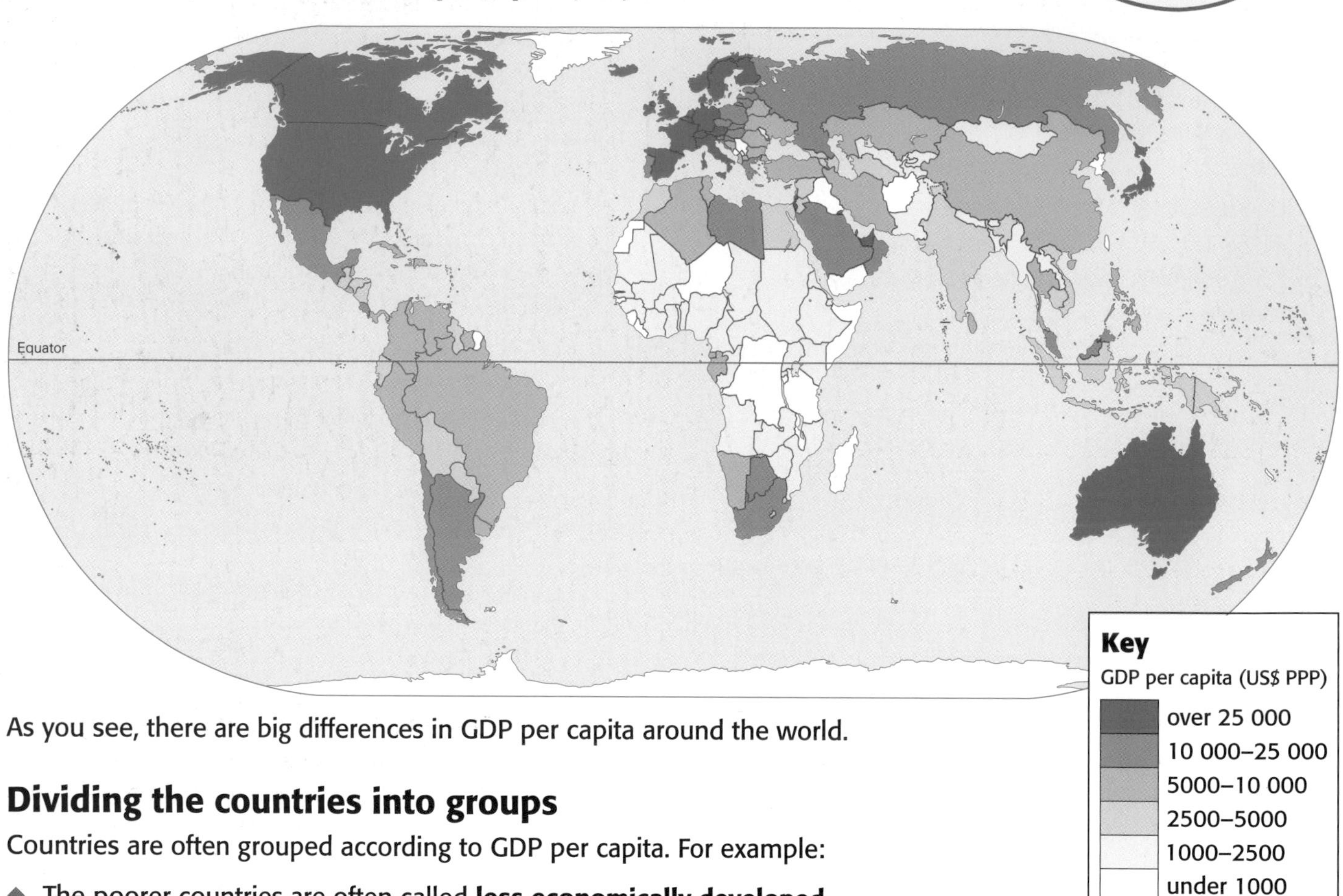

As you see, there are big differences in GDP per capita around the world.

Dividing the countries into groups

Countries are often grouped according to GDP per capita. For example:

- The poorer countries are often called **less economically developed countries** or **LEDCs**.
- The richer countries are called **MEDCs**. (What does that stand for?)
- The poorest countries are also called the **Third World**. But many geographers don't like this term – they think it's patronising.
- The richer countries are sometimes called the **rich north**, because most are in the northern hemisphere. So the poorer countries are called the **poor south**.

But it is always changing …

Not so long ago, countries like India, China, Brazil, and South Korea were counted as poor. But they are developing fast. This table shows how fast China's GDP per capita is growing. (Chapter 2 has lots more about China.)

These countries are developing fast by setting up industries. So they are called **newly industrialised countries** or **NICs**. (The UK and some other European countries began to industrialise over 200 years ago.)

GDP per capita for China

Year	GDP per capita (US dollars PPP)
2000	$4000
2001	$4020
2002	$4580
2003	$5000
2004	$5900
2005	$6800
2006	$7700
2007	$8900

More about LEDCs

**Some poorer countries (LEDCs) are large. Some are small.
But they do tend to share some features.**

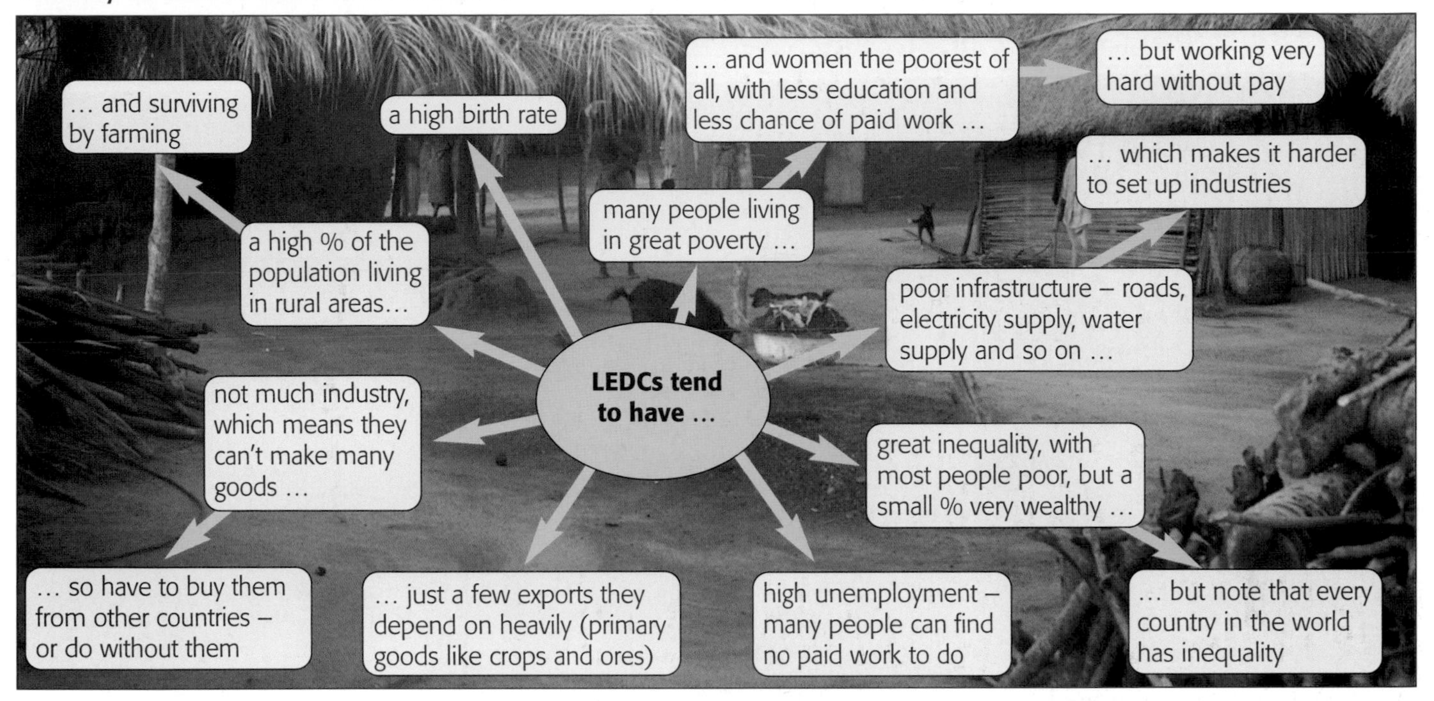

**Most LEDCs are striving hard to develop, and give their people better lives.
So a map of GDP per capita should look different 20 years from now.
And note that even the poorest countries have some wealthy people, and better-off areas.**

Your turn

The map on pages 140–141 will help for some of these.

1 Look at the map on page 16. In which range of GDP per capita (in US dollars PPP) is:
  a the UK? **b** Ghana? **c** Brazil? **d** Japan?

2 Name:
  a five other countries in the same group as Ghana
  b five other countries in the same group as Brazil
  c five of the world's very richest countries for GDP per capita.

3 Assume for now that the MEDCs have a GDP per capita (PPP) of $10 000 or over.
  a What does *MEDC* stand for?
  b Name two MEDCs you haven't named already.

4 Write out each sentence. After it, write *True* or *False*.
  A Overall, Africa is the poorest continent.
  B Iceland is in the highest income group.
  C Mali is one of the world's poorest countries.
  D Everyone in Mali is really poor.
  E Overall, Libyans are better off than Egyptians.
  F The GDP per capita for Japan is $20 000 (PPP).

5 If you map *life expectancy* on a world map, you will get a pattern very like the one on page 16.
See if you can draw a diagram like the one started below, to explain why. Add more boxes and arrows!

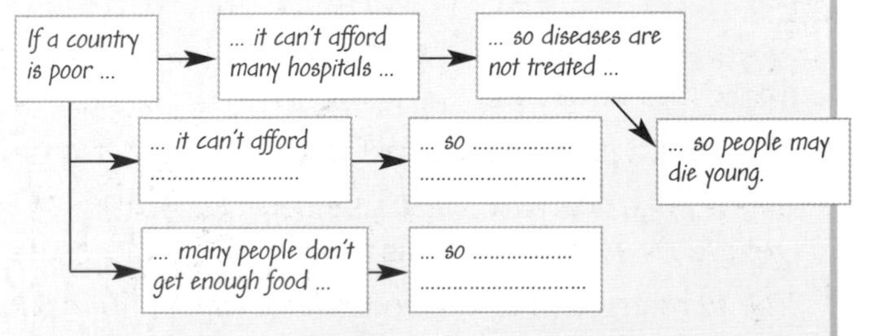

6 Find Cambodia on the map on page 16. Then, using the spider map above to help you, write 6 bullet points about development in Cambodia. Start like this:
I think Cambodia is likely to have …

7 *Why* are so many countries poor? Give all the reasons you can think of. (But no silly ones!)
For example could a lack of natural resources (such as good farmland) be a reason? What about climate? Could other countries be to blame, in any way?

How did the development gap grow?

This unit is about reasons for the differences in development around the world.

Some reasons for the development gap

1 Historical reasons

2000 years ago, India and China were the world's richest countries. But over the centuries, the pattern changed. Countries in Europe and North America raced ahead. Let's see how that happened.

By 1500 AD, Western Europe had begun to develop fast, thanks to good farmland, a mild climate for farming, and industries such as shipbuilding and textiles.

Then the Industrial Revolution came along. It began around 1750 in Britain and spread to Europe. It led to a leap in wealth and development for several countries.

Europeans had already settled in North America by then. So they started to develop industries too. And soon North America began to grow wealthy.

Meanwhile, Europeans had been exploring Africa, South America, and Asia. They had found lands rich in natural resources. Trading soon followed.

It began with friendly trading for things like gold, tobacco, timber, and spices. The Europeans gave goods in exchange. But as time went by they grew greedier, and ...

... took over many of their trading partners by force, as **colonies**. They took their raw materials, and sold them finished goods – and that made many Europeans very rich !

In time, the Europeans were forced out. They left behind them countries with very little industry, low levels of education and skill, and often a great deal of unrest.

Between them, countries like Britain, France, Spain, and Portugal carved up Africa, South America and much of Asia as colonies. Many of their ex-colonies are still poor today – and some are still unstable.

2 Geographical reasons

A country's location, and climate, and natural resources, can play a huge part in helping it to develop.

Its coal, oil, gas, and good farmland have all helped the UK to develop. And the sea has been great for fishing, and trading.

But in a hot dry country far from the sea, isolated by mountains, with poor soil and few other resources, development may be very difficult.

Some countries have the opposite problem – good soil, but too much rain, and severe floods. Years of hard work get washed away.

3 Social and political reasons

A country has a better chance of developing if it is stable and secure, with a strong government.

But many of the world's poor countries have wars going on, with a big waste of lives, and money.

And in many countries, corrupt leaders have made themselves rich, while their people live in poverty.

Your turn

1 **A – I** are facts about some different countries. For each, explain why this could have held back development.
  - **A** It is mountainous and hard to reach.
  - **B** A tribal war has been going on there for 10 years.
  - **C** Millions of its people are suffering from AIDS.
  - **D** It suffers severe flooding almost every year.
  - **E** It was a British colony for over 50 years.
  - **F** A small group of people owns most of its wealth.
  - **G** Others refuse to trade with it, because of its politics.
  - **H** It has poor soil, and the rains are not dependable.

2 Look at the facts in **1**. Which of them are:
  - **a** historical (about things that happened in the past)?
  - **b** geographical?
  - **c** to do with society and politics?

3 Of all the conditions described in **1**, which ones do you think could be put right, or at least improved, to help that country develop? Explain each choice you make.

4 The UK is among the world's most developed countries. See if you can give 8 reasons to explain why. (At least 2 geographical, 2 historical, and 2 social/political.)

These don't help either

Here we look at some other factors that help to keep the development gap wide.

Keeping the gap wide

In the last unit, you saw some reasons for the big gap in development around the world. Now we look at some other factors that help to keep a poor country poor – no matter how hard it works.

Big debts to pay off

Many poor countries have been paying out millions of dollars a year, as interest on money they borrowed. This is how it happened:

Naturally, poor countries want to develop fast – but that needs money. So they have borrowed lots of money. Some from ...

... ordinary **banks** like we all use, some from the **World Bank** (a special bank that countries set up to help each other) ...

... and some from other **governments**. To get the money, the poor countries often had to make promises in return.

Here's the sums ...

LOAN: $1000 MILLION

INTEREST RATE: 2% PER YEAR

2% OF $1000 MILLION = $20 MILLION

SO EACH YEAR WE PAY INTEREST OF 20 MILLION DOLLARS!!!!

When you borrow money, you pay interest on the loan. Many bank loans were made in the early 1970s when interest rates were low.

Then they shot up. So the poor countries had to use more and more of the money they earned, just for interest payments.

That meant less money for schools, and hospitals, and a water supply, and the other things their people badly needed.

It became clear that this was a bad situation. So the World Bank, and governments of richer countries, agreed to cut interest payments for many **heavily indebted** countries.

Some debts have also been cancelled, for the countries in most debt. For example in 2004, $2.6 billion of Ghana's debts were cancelled. Ghana still owes $2.4 billion. But now it pays out less on interest each year. (It still needs to borrow new money for future development!)

Trading troubles

Many poor countries depend on selling crops such as sugar, cocoa, and coffee, to other countries. These are crops we all want. So why are those countries not getting a lot richer? Let's see.

When the poorer countries try to sell their crops to richer countries, they often face big import taxes or **tariffs**. These put buyers off.

At the same time, the world price for many crops has been falling over the years. Partly because too much is being grown …

… and partly because the big food companies, who buy up most of the crops, are so powerful that they can force the price down.

The tariffs mean the poor countries can't sell so much. Falling world prices mean they earn less from what they do sell. But their problems don't end there …

… because meanwhile, farmers in rich countries grow many of the same or similar crops – and get grants or **subsidies** for doing so.

These crops are then exported to the poorer countries. At prices so low that the local farmers can't compete, and go out of business.

The poorer countries can't stop these imports, because the World Bank has forced them to reduce or drop tariffs, in exchange for loans.

So poor countries that depend on exporting crops may stay poor, no matter how hard people work.

What can be done to help poor countries catch up, and close the development gap? Unit 1.10 has some ideas.

Your turn

1. Explain what these terms mean. (Glossary?)
   a the World Bank **b** debt **c** interest rate
   d tariff **e** subsidy
2. You live in a poor country. Your government wants to borrow money from a rich country, to build schools – but wants your advice first. What will you say?
3. Many countries depend on exporting crops – and they remain poor, no matter how hard they work. Give reasons.
4. You are a farmer in a poor country. You grow rice, and keep chickens. Your country has been forced to drop tariffs on imported rice and frozen chicken. Explain why this will: **a** harm you **b** help farmers in other countries

So why is Ghana an LEDC?

In the last two units, you met reasons why many countries are less developed. Here you'll find out how those reasons apply to Ghana.

Working hard – but still poor

Ghana has many natural resources, including gold and diamonds. It's the world's second largest producer of cocoa. Its people work hard. But millions of them still live in deep poverty. Why ?

▲ *A cocoa farmer. The yellow pods contain the cocoa beans, that end up in chocolate.*

1 Some historical reasons

We were not always one country. Once we were separate kingdoms and tribes, with our own languages and culture. And some of us were wealthy.

Then came the Europeans. First the Portuguese, in 1650. Later the Dutch, Danish, Germans, British. They were so excited by our gold. So we traded it. The British called us the Gold Coast.

By 1650, they wanted slaves more than gold, for their plantations in the Americas. They bought at least 5000 people a year from us. We had some slavery before – of our enemies, and people to be punished. But now tribes fought with each other to get people to sell. Villages and families were torn apart. It went on for over 150 years. So shameful.

The Europeans competed to trade with us. They set tribes against each other. But Britain took over, little by little. By 1901 our lands had become a British colony. Our kingdoms and tribes were forced together.

The British shipped out gold, diamonds, ivory, pepper, timber, corn, cocoa. They built railways to carry them to the coast. They did build some roads and schools and hospitals too – but made us pay for these through taxes.

In the end, we had enough. We wanted freedom ! At last, in 1957, we gained independence. We called our country Ghana, after an ancient West African kingdom. Free – but with no factories, few services, and few skilled people. And an uneasy relationship between the tribes Britain had forced together.

We have been free for less than 60 years. That's not very long, is it?

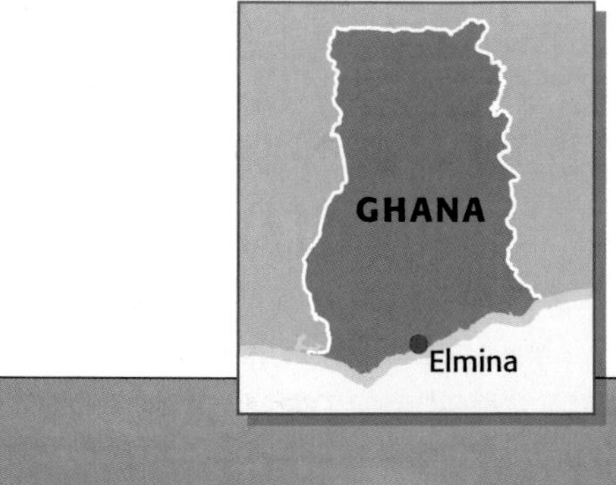

▲ *The fort at Elmina, where slaves were chained in dungeons until the next slave ship arrived.*

2 Some environmental reasons

56% of Ghana's workforce are farmers. So land is very important to them. But in the north, land is turning into desert. This is called **desertification**. It is due to drought, chopping down trees, and overgrazing.

In the south, three-quarters of the rainforest has been destroyed – for timber, firewood, and land to grow cocoa. This is called **deforestation**. People gain more land to farm, but the exposed soil is soon useless. And many can't afford fertiliser to improve it.

All this means smaller crops to eat and sell. Which means more poverty.

3 Some political reasons

Since independence, Ghana has suffered much political unrest. The army took over the country for three separate periods. But now it is stable, under an elected government. People say it will remain stable.

4 Big debts to pay off

Ghana borrowed a lot of money in the past. So it had to pay out lots of **interest** each year – which meant less money for development.

In 2004, Ghana was given **debt relief**: richer countries and the World Bank agreed to cancel over half its debts. But it still owes $2.6 billion. And it still needs to borrow more to help it develop.

5 Trading troubles

- Ghana relies heavily on cocoa, to earn money. But the world price of cocoa goes up and down. So, some years, Ghana earns much less from it. (And most goods that Ghana imports get more expensive.)
- Many Ghanaian farmers grow rice, for sale in Ghana. But far cheaper rice is coming in from Europe, Asia and the USA, where rice farmers get subsidies. This has put many of Ghana's rice farmers out of business.
- In the same way, Ghana's chicken farmers can't compete against the frozen chicken coming in from Europe and the USA.
- In exchange for debt relief, Ghana is not allowed to stop these imports. Many people think this is unfair. They think poor countries should be allowed to protect their farmers and fragile industries.

But Ghana hopes that its new-found oil and gas will help to solve its problems, and turn it into an MEDC.

▲ *A farm in northern Ghana, at risk of desertification.*

Your turn

1 You have to draw a time line for Ghana.
 a On a large sheet of paper draw a vertical time line from 1450 up to 2010. Make it 30 cm long if you can. (Use two pages ?)
 b On your line mark in events from both the text **and** the box below. (Small neat writing !) Add a title.

Year
1500
1450

2 Beside your time line shade the period in which:
 a West African slaves were bought by Europeans
 b the Gold Coast was partly or wholly a British colony.

3 Now underline the events that you think:
 a *helped (or will help)* Ghana to develop, in one colour
 b *held back* its development, in another colour
 c *did a mixture of both*, in a third colour.

4 Add a key for your colours and shading, for **2** and **3**.

5 Choose *one* event you underlined for **3c** above, and explain why you underlined it.

6 *'Since independence, Ghana's development has been completely under its own control'.*
From the work you have just done, do you think this statement is true ? Give your reasons.

EVENTS

- 1878: a Ghanaian brings back cocoa plants from Fernado Po, an island off Africa
- 1528: chocolate drink from the Aztecs introduced to Europe, by Spanish explorers
- 1817: slavery abolished in Europe
- 1657: London's first drinking chocolate café opens
- 1928: a large harbour built at Takoradi
- 1502: first slave ship leaves West Africa
- 1965: the Akosombo dam completed, to provide Ghana with hydroelectricity
- 1980: economy almost collapses due to low cocoa price and other problems
- 2004: over $2 billion of Ghana's debts are cancelled
- 1874: Britain takes control of the south of the Gold Coast
- 1885: the first cocoa exported to Britain
- 1618: first British trading settlement set up on the Gold Coast
- 1999: crisis in Asia and Russia causes world chocolate sales to fall
- 1993: Ghana earns $222 million from selling rainforest timber
- 2007: Large deposits of oil and gas found off Ghana's coast
- 1830: the world's first chocolate bars made in England by J S Fry and Sons
- 1898–1927: railways built by the British
- 1983: Ghana has to pay back loans of $1.5 billion to other countries

Tackling the development gap

This unit is about ways to try to close the big development gap between countries.

How can we close the development gap ?

Closing the gap between rich and poor countries needs money! To build the schools, and hospitals, and roads, and other things people need.

Poorer countries are working hard to earn the money for themselves. But there are also things rich countries can do to help them.

Did you know?

◆ The world has enough money, and know-how, to end poverty.

1 Poor countries helping themselves

1 Using their natural resources
Most LEDCs depend on one key resource: soil. They are working to improve their crops, and try new crops to sell. (Such as flowers.)

Some have natural resources which they have not yet exploited. Such as oil, and gas, and metal ores. These could bring in a great deal of money for development.

2 Providing services
For many LEDCs, tourism is a good way to earn money. People visit and enjoy themselves. (But Chapter 6 gives some pitfalls!)

Some LEDCs offer business services, such as call centres, and computing services. Companies in MEDCs use them to save money.
(Wages in LEDCs are lower.)

3 Manufacturing
Manufactured goods can earn you more than crops do. So LEDCs are keen to set up factories, making things to export (and sell at home).

For example clothing factories are a good start. Everyone needs clothing. It can't be sewn by robots: you need to employ lots of people. And the skills are easy to learn.

But poorer countries may need a little help, to get started.
For example drilling oil wells costs a lot, and needs special expertise.
So big foreign companies may offer to help, in exchange for a share of the profits. The LEDCs just need to make sure it's a fair deal!

2 Richer countries helping

Richer countries can do a lot to close the development gap. For example:

% of GNI given as aid in 2005

Country	%
Australia	0.25
Canada	0.42
Denmark	0.81
France	0.47
Germany	0.36
Japan	0.28
Netherlands	0.82
Norway	0.94
United Kingdom	0.47
United States	0.22

They promised 0.7%.

1 Cancel *all* old debt
This would free the LEDCs from a big burden, and help them to make a fresh start.

2 Give more aid
Rich countries often promise lots of aid to poor countries – but then break their promises.

3 Make world trade fairer
This would help LEDCs to earn more, which they could then use for their own development.

The Millennium Development Goals

In 2000, the world's countries agreed to make a big effort to tackle the development gap. Here are just four of the goals they set for 2015:

The Millennium Development Goals

By 2015 we aim to:

- *halve the % of people living on less than 1 dollar a day*
- *halve the % of people without access to a safe water supply*
- *cut under-5 deaths by two-thirds*
- *ensure that all children, everywhere, complete primary school*

Each LEDC then set its own goals, to help meet the overall goals. MEDCs have helped by cancelling some debts, and giving grants and other aid. Some countries, like India and China, are likely to meet their goals. But many African countries have fallen behind. And 2015 draws near.

Your turn

1 a You have learned quite a lot about Ghana. Now see if you can suggest ideas for new factories and industries for it. Try for at least three. (Page 11 may help.)

b Explain how these could help Ghana to develop.

2 a Look at those Millennium Development Goals. Why don't they aim for safe water for *everyone* by 2015?

b Arrange the four goals in what you think is their order of importance (most important first), and explain why you chose this order.

3 Meeting the Millennium Development Goals will cost a fortune. See if you can explain why.

4 The little table at the top of the page is about aid.

a What is *GNI*? (Glossary?)

b How many of those countries kept their aid promise?

c Write a speech to make to the leaders of the other countries, saying why they should keep their promise.

5 Could *you* do anything to help a poor country to develop? Explain.

1.11 Small is beautiful

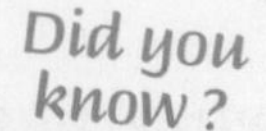

Did you know?

◆ About 1100 million people around the world do not have access to a supply of clean safe water.

Development is not just about big expensive projects like dams and airports. Here you'll see how a small local project can improve people's lives.

Ghana's water problem

◀ *Like a drink of this?*

This is Lamisi. And this is her family's water supply, for drinking, cooking and washing. She has been here collecting water for over three hours.

The water in the bucket looks very muddy. But far worse than the mud are the things you can't see: bacteria that cause diarrhoea, typhoid, and cholera; and tiny eggs that grow into worms inside you, leading to bilharzia and other diseases.

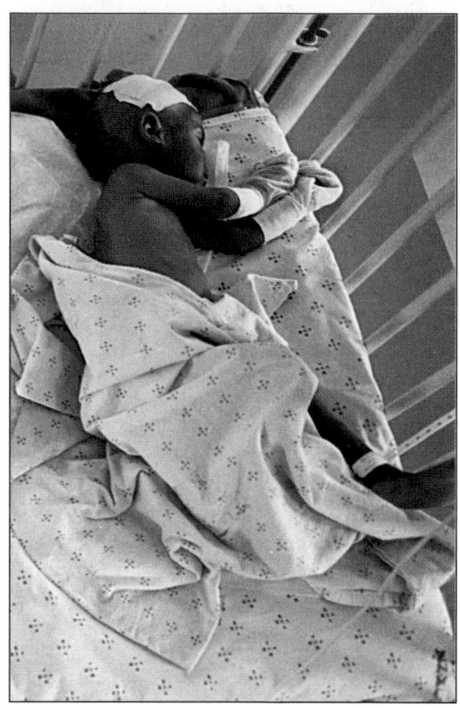

▲ *This baby has cholera, one of the many diseases caught from dirty water.*

Development little by little

Lamisi is not alone. Over 5 million Ghanaians have no access to clean safe water. They get their water from rivers and ponds.

One day everyone in Ghana will have piped water. But that could still be years away. People can't wait. So, right now, many villages are digging wells for themselves, with help from a British charity called WaterAid.

Everyone in the village gets involved:

WaterAid supplies the know-how, the materials for lining the well, and the pump.

↓

Villagers form a committee to decide where the well will be, and organise the work.

↓

Everyone in the village joins in to help clear the site, and dig, and carry soil away.

↓

Some villagers are trained to look after the well and carry out repairs.

Cost of a hand-dug well: about £1200.

▲ *A new well. Everyone helped to build it, and everyone benefits.*

The difference a well makes

This is Abena and her baby. Abena and other villagers were asked how the new well had helped them. This list shows their answers. So wells don't just bring clean water!

The changes we noticed

- ☑ A more young people have time to go to school
- ☑ B teachers happier to stay in the villages to teach
- ☑ C much less illness, so less spent on medicine
- ☑ D women potters can produce more pots
- ☑ E more people cooking food to sell
- ☑ F more people selling iced water
- ☑ G no more quarrels with neighbouring villages about water
- ☑ H people take more pride in the village
- ☑ I cooked foods look much better
- ☑ J visitors can be offered clean drinking water
- ☑ K clothing and homes kept cleaner
- ☑ L much less time taken to fetch water
- ☑ M less far to walk for water, so less tired

Your turn

1 Suppose Lamisi had a choice: electricity for her village, or a well and pump like the one in the photo. Which do you think she'd choose? Why?

2 This diagram shows a hand-dug well, and pump.

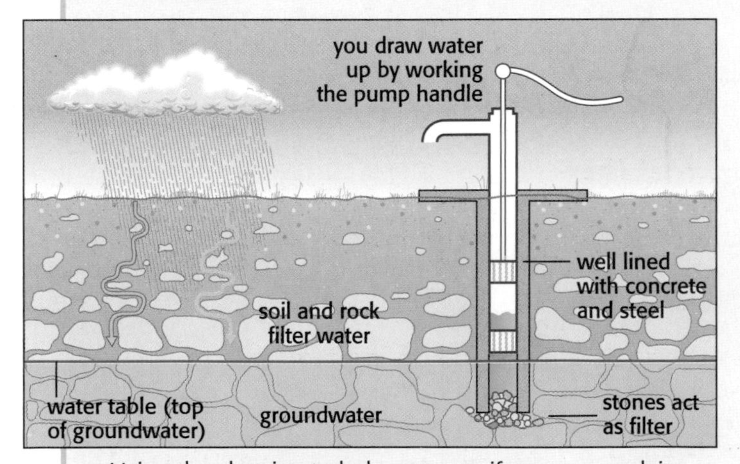

Using the drawing to help you, see if you can explain:

a what groundwater is
b where it comes from
c why the water that gets pumped up is clean
d why the wells don't cost all that much
e why villagers need a bit of help from experts

3 Now look at the list of changes **A** – **M** above.

a Draw a larger copy of the Venn diagram on the right.
b Write letters **A** – **M** where you think they should go. At least one where the loops overlap !
c Choose one change that you placed in the overlap, and explain why it belongs there.

4 Now look at Abena's toddler above. She is called Aba. What effect might the well have had on Aba's life:

a by a year from now?
b by twenty years from now?

5 This is one person's opinion. Do you agree with it? Write down what you would say in response.

Digging wells is a waste of time. It does not solve the problem of poverty.

6 **a** Look at the title of this unit. What does it mean ?
b See if you can think of other ideas for small projects that could make a big difference to Abena's village. (It's like Grace's village on page 13.)
c i Choose the one you think the villagers would like most.
ii How much do you think it would cost? (Guess!)
iii Write an action plan for what you would do, to raise the money for it.

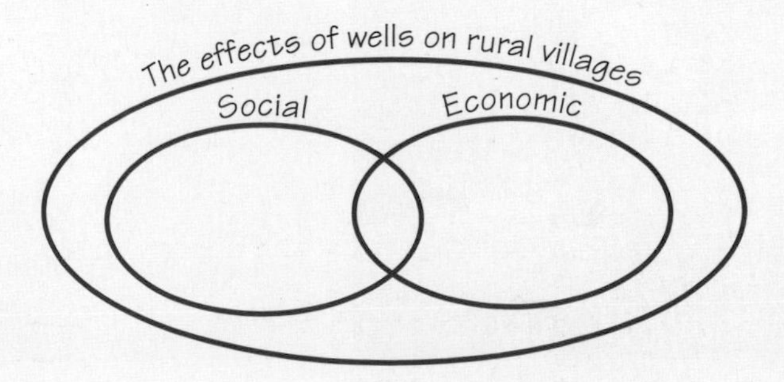

2 Close-up on China

Leave the UK, and head eastwards. About 8000 km later, you'll arrive in the People's Republic of China! Where you will find ...

▲ *... a famous wall ...*

▲ *... huge cities (this one is Shanghai) ...*

▲ *... farmers hard at work...*

▲ *... hundreds of thousands of factories, making things we buy in the UK...*

▲ *... delicious food...*

▲ *... exotic wildlife...*

▲ *... and magical dancing dragons, to bring you good luck.*

The big picture

This chapter is about China. These are the big ideas behind the chapter:

- China has one-fifth of the world's population.
- It is developing fast, thanks to industrialisation and trade.
- Its development has lifted hundreds of millions of Chinese people out of poverty.
- But almost one-tenth of China's people still live in great poverty.
- China's impact on the world is growing, as it develops.

Your goals for this chapter

By the end of this chapter you should be able to answer these questions:

- Where in the world is China? (Give at least four facts about its location.)
- What can I say about its relief, and other physical features? (At least five facts.)
- What can I say about its climate? (At least four facts.)
- How is its population spread around, and what factors have influenced the pattern? (Try for three factors.)
- Which Chinese cities can I name, and place on a map? (At least four.)
- What is the one-child policy, and why was it introduced?
- China's GDP has been rising fast. Why is this?
- Inequality has been growing, in China. Why?
- Millions of people have been leaving China's rural areas. Why? Where have they gone?
- China has some severe environmental problems to tackle. What examples can I give? (At least two.)
- Where is the Three Gorges dam, and why was it built?
- China's impact on the world is growing. What examples can I give? (At least five different examples.)

And then ...

When you finish this chapter you can come back to this page and see if you have met your goals!

Invented in China!

Invention	In use in China by ...	Being made in Europe by...
wheelbarrow	231 BC	1200*
paper	105	1120
gunpowder	800	1300
porcelain	around 800	1709
printed book	868	1456
compass	1040	1190

* The years are AD unless BC is stated.

Did you know?
- The Great Wall of China was over 6200 km long – the largest human structure in the world.
- A lot of it has gone now.

Did you know?
- Red is counted a lucky colour, in China.

Did you know?
- China's population grows by about 1 million people every 5 weeks.

Your chapter starter

The photos on page 28 show China.

Where is China?

How much do you know about it?

Which photo fits best with your own mental image of China?

Did any of the photos surprise you? Which?

2.1 China's physical geography

Find out here about China's main physical features, and its climate.

It's big!

China is big. It's the fourth largest country in the world, after Russia, Canada, and the USA. It's just a little smaller than the USA – and about 40 times the size of the UK!

A

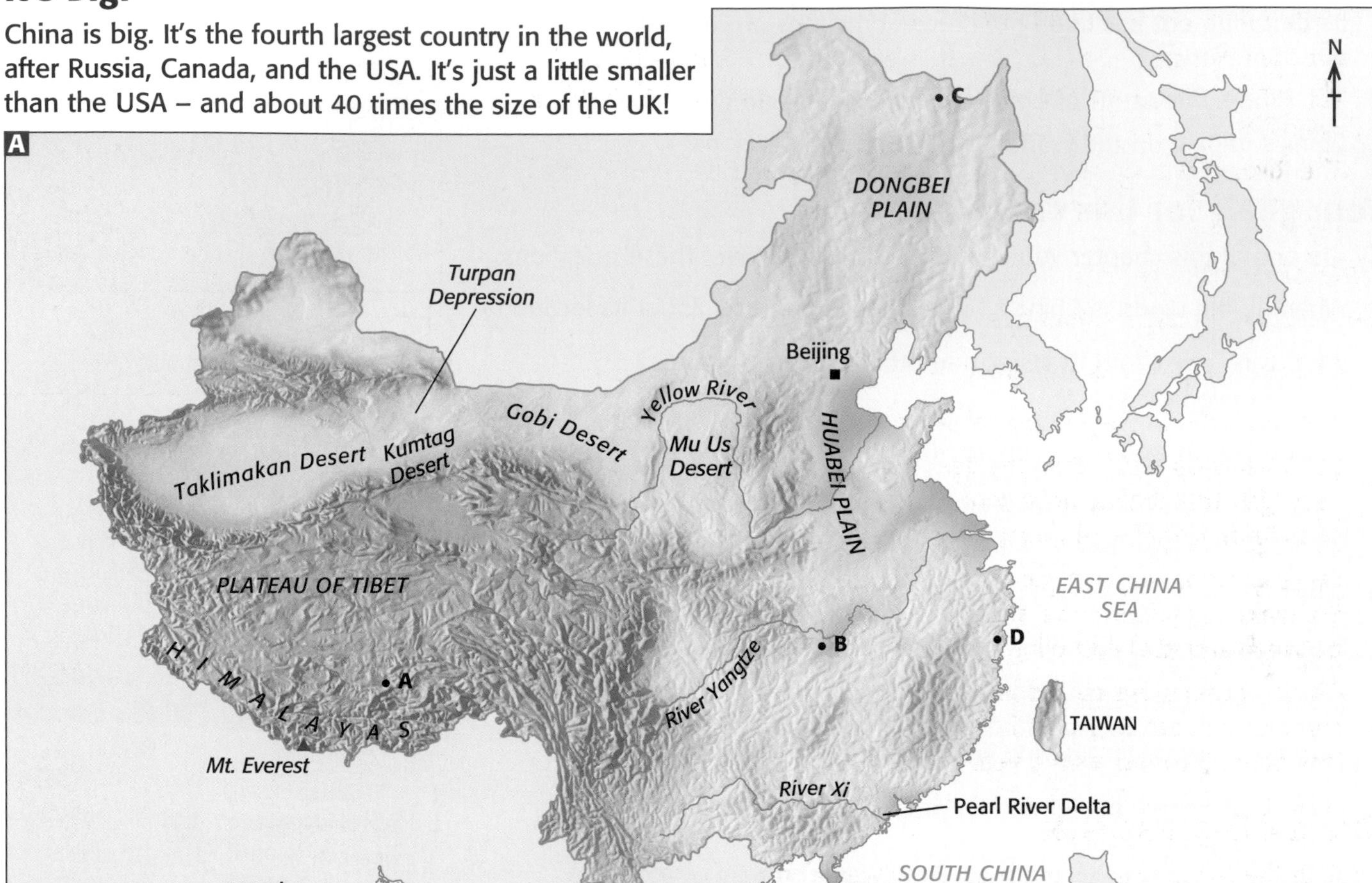

Relief

As you go west in China, the land steps upwards. Look at the map.

- The lowest areas are along the coast. The Huabei Plain (or North China Plain) is the largest area of flat land. It is very fertile.
- The next step up has mountain ranges, and large deserts. Almost 20% of China is desert. Some of the deserts are sandy, and some stony.
- The top step has the vast Plateau of Tibet. It is really high here – about 4000 m above sea level, on average. The Himalayas form its southern border. And note Mt Everest. It lies on the border with Nepal.

▲ *This man wades through floods, after a storm in south east China …*

◀ *… while this man wades through the Kumtag Desert in north west China. He's taking his camels to a tourist spot!*

Rivers

China has thousands of rivers. The map shows just three. Note that:

- the Yellow River is the world's sixth longest river. It is named after the yellow silt it carries. This is deposited in the lower course of the river, raising the bed – and causing heavy flooding.
- the River Yangtze is the world's third longest river. It is really busy – like a motorway for ships and boats. It too has a history of flooding. Read about its big new dam in Unit 2.9.
- The River Xi is also very busy. It flows to the Pearl River Delta, which is one of China's top industrial areas.

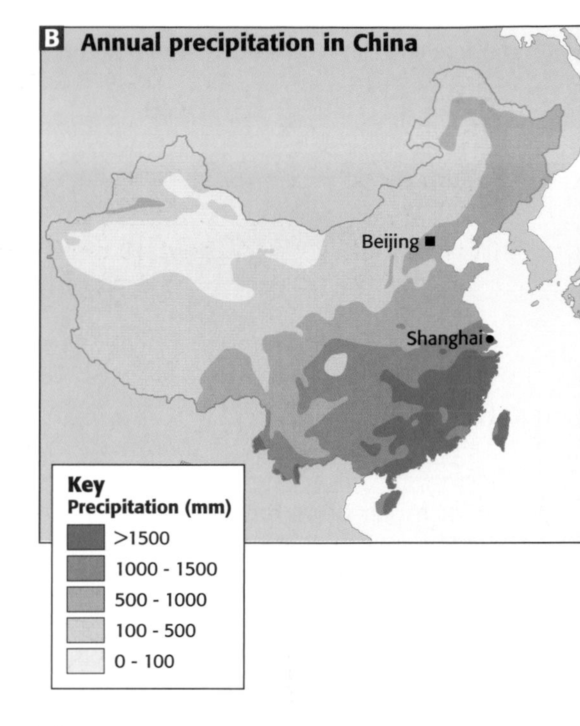

B Annual precipitation in China

The climate

Because it is so large, China has a range of climates.

- The north of China is sub-arctic. The far south is in the tropics.
- It is very cold on the Plateau of Tibet, since this is so high.
- Land heats up faster than the sea in summer. It heats the air, which rises fast. This draws moist monsoon winds in from the sea, bringing plenty of monsoon rain from the south east.
- Large land masses heat up fast in summer, and cool fast in winter. So inland, away from the sea, you find big temperature differences between summer and winter.
- China's deserts are **cold deserts**. They may get warm or hot in summer, but are very cold in winter, and at night. (Any place with less than 25 mm of rain a year counts as a desert, even if it's cold.)

C China's climate regions

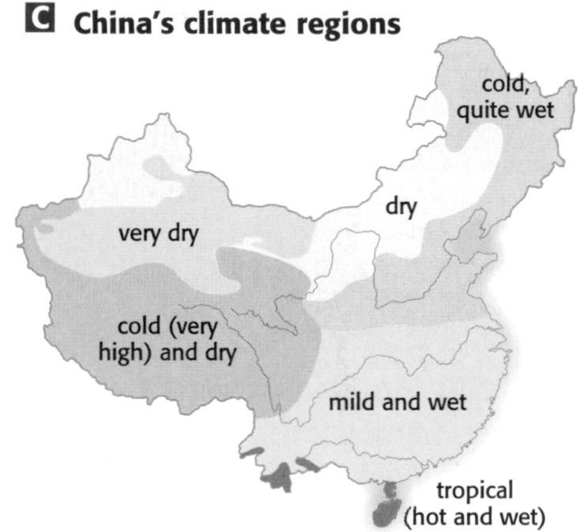

The earthquake risk

Look at the Himalayas on the map. They were formed by two plates (the Eurasian plate and Indo-Australian plate) pushing into each other.

The plates are still pushing, so the Himalayas are still growing, by over 1 cm a year! The plate movements cause many earthquakes in China. In 1976, a major earthquake killed over 250 000 people.

Your turn

1 First, where is China? Give five facts about its location in the world. (For example, which continent is it in?) The map on pages 140 – 141 will help you.

2 a What does *relief* mean?
b See if you can describe the pattern of relief in China in just 25 words.

3 Look at places **A** – **E** marked on map **A** on page 30. See if you can explain why:
a it is always cold at **A**
b it is much warmer at **B** than at **C**
c the temperature difference between summer and winter is greater at **B** than at **D**
d you'll find palm-fringed tropical beaches at **E** (on Hainan island)

4 a In China, most rain falls between May and October, in the *sonomon* season. Unjumble the word in italics!
b What causes the rain, during this season?

5 Look at map **B** above.
a Which region gets most rain? Choose one:
north west *north east* *south east* *central*
b Explain *why* it gets the most rain.
c i Now choose the driest region, from the list in **a**.
ii Try to explain *why* it is driest. Map **A** might help.
d Where might you expect snow rather than rain?

6 Look at the Kumtag desert, in the photo and on the map.
a What makes it a desert?
b It's a different type of desert than the Sahara, in one key way. Which way?

2.2 A little history

Here you can read a brief history of China. It will help you see how China is changing.

The Chinese Empire

For much of its recorded history, China has been ruled by Emperors. The first was Qin Shi Huangdi, who became Emperor in 222 BC. (The Great Wall of China was his idea.)

The title of Emperor passed from father to son. But rebellions and invasions led to changes in China's dynasties, and borders, over the centuries.

China developed into a wealthy and civilised nation. It held science, philosophy, and the arts in high esteem. The Chinese came up with many important inventions.

By 1800, European traders were buying many goods from China. But China did not want European goods in return. It just wanted payment in silver.

So British traders began to sell people opium. China tried to stop this. It led to the first Opium War between Britain and China (1840 – 1842). China was defeated.

Other countries then forced China to sign trading treaties. Later, Britain and France joined forces to fight a second Opium War with it (1856 – 1860). China lost again.

China was left weak. Some Chinese feared it would be taken over. They set up the Nationalist Party. Support for this grew, and in 1912 the Emperor was forced to resign.

But the Nationalists were not able to solve many of China's problems. A second group, the Communists, struggled with them for power – and eventually won.

In 1949, the Nationalist leaders fled to Taiwan, an island off China, and set up a government there. The Communist Party took control of China – and is still in control today.

Today, Taiwan still has its own government. But China insists that Taiwan belongs to China. So the relationship between the two is uneasy.

The People's Republic of China

When the Communist Party took over, it aimed to make China strong and self-sufficient, with everyone equal. The leader was called Mao Zedong. He named China the **People's Republic of China**.

Before that, people could own land. Under Chairman Mao, no individual could own land. Instead, it was shared by large groups or **communes** of farmers. They also shared the work between them.

The state took over factories and other businesses too. People were told what work to do. They had to ask permission to move to other places. The usual answer was no. (Mao did not want everyone moving to the cities.)

In return for their work, people had free food, and education, and health care, and child care, and support when they grew old.

But overall, it was not a success. In the period 1958 – 1961, not enough food was grown to feed China. Over 20 million people died of famine.

▲ *A protestor carries a portrait of Mao Zedong, in 2008. His name is also written as Mao Tse-Tung. (Mao is his surname.) The protest is against the opening of a French supermarket!*

The Cultural Revolution

In 1966, Mao launched the **Cultural Revolution**. His aim was to wipe out the Four Olds: Old Customs, Old Culture, Old Habits, and Old Ideas. Millions of **Red Guards** did most of the work for him. These were mostly young people. Many were students. Many were your age.

The revolution lasted over three years. It was a time of chaos. Old temples and shrines, museums, libraries, statues, and paintings were destroyed. Religion was banned. The Red Guards spied on their own parents and neighbours, and reported on them. At least a million people were killed. Millions more were sent to labour camps.

China today

Mao Zedong died in 1976. He had changed China. But he had failed to turn it into a modern industrial nation – and the Cultural Revolution was a disaster.

So in 1979, economic reform began. Now, farmers lease plots of land to farm. They must sell some produce to the state, but can sell the rest to anyone, and keep the money. People can set up their own businesses. Chinese companies can team up with foreign ones.

As a result, China is developing rapidly. The standard of living is rising fast. You can find out more in Unit 2.5.

Communism

These are the main ideas behind pure communism:

- Wealthy people who own factories and land use other people as workers, just to make profits.
- So workers are like slaves in some ways (except they get paid).
- But we are all equal.
- So, nobody should be allowed to own property. The state should own everything.
- The government can plan what to grow and make, so that everyone has what they need.
- Then people work to produce these things …
- … and in return, get everything they need, for free.

Your turn

1 China has had many different dynasties, in its history. What is a *dynasty*? (Glossary.)

2 a What caused the first Opium War?

b Some say the Opium Wars are a shameful blot on Britain's history. What do you think?

3 Look at the panel above, about communism. Do you think this system has good points? Explain.

4 What might *you* find difficult, about living in a communist country? Give reasons for your answer.

5 After Mao's death, many policies changed, in China. Now people can get permission to set up businesses, and make profit for themselves. This is helping China to develop. See if you can explain *why*.

2.3 The world's largest population

Here you'll see how the population of China is spread around, and explore reasons for the pattern.

▲ *China has many ethnic groups. But most Chinese people are Han.*

Tops for population

About a fifth of all humans live in China! It has the world's largest population: over 1.3 billion people. (That's 1300 million.)

Where do they live?

This map shows how the population is distributed.
The deeper the shade in an area, the more people live there.

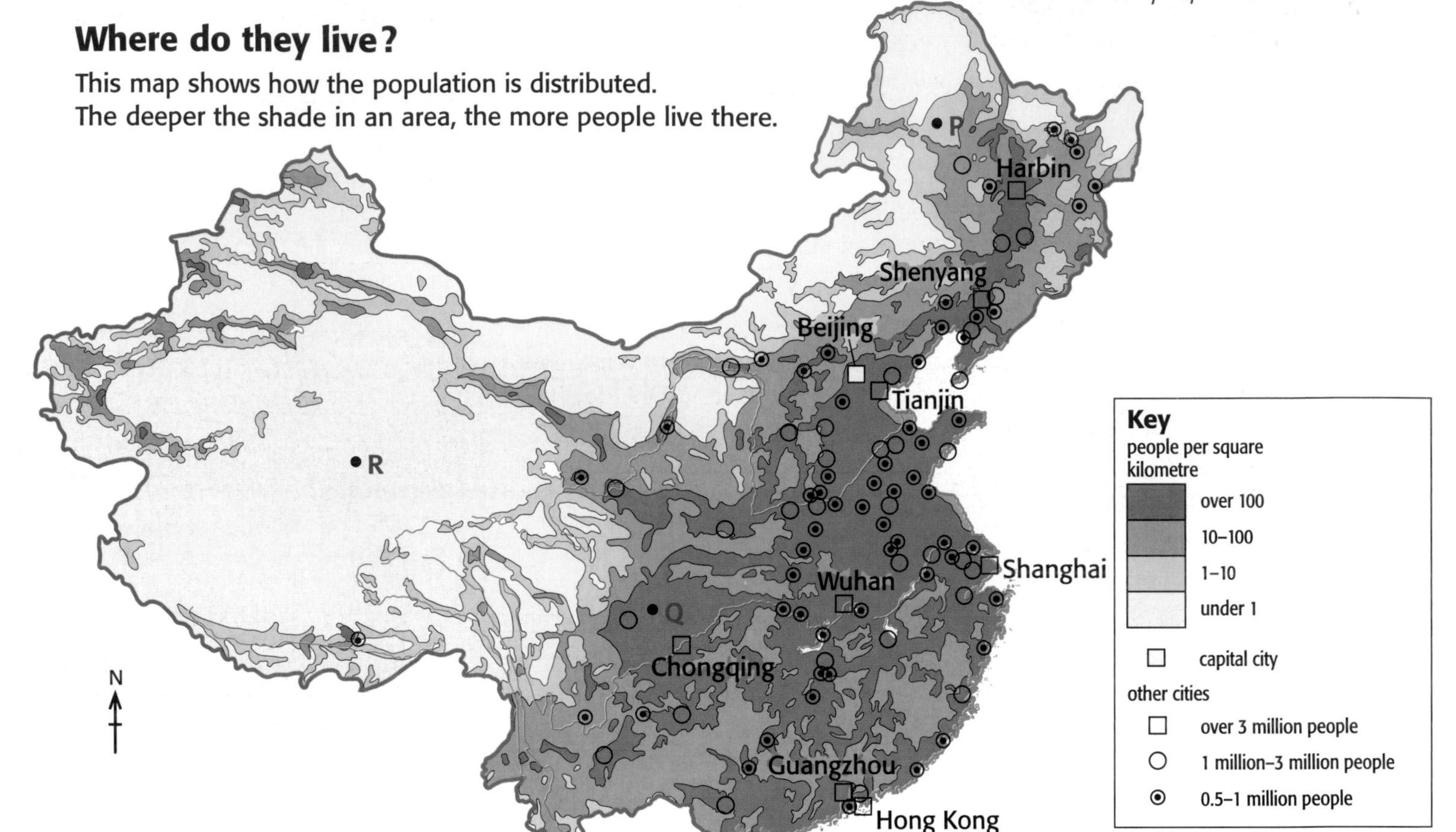

Overall, around 95% of the population lives in less than half of China's area. Large parts of the country are empty, or very sparsely populated.

China's big cities

Around 44% of China's population lives in towns and cities. The map shows just the largest cities. China has around 650 cities altogether.

- Beijing is the capital city. (It used to be called Peking.)
- Shanghai is China's largest city.
- Hong Kong has a population of 7 million, but covers only a small area. It is one of the most densely populated places in the world.
- 55 of China's cities have over 1 million people. Many cities have large built-up areas around them, with hundreds of thousands more people.

Towns and cities have been growing fast, in China. A big reason is that people have been arriving from rural areas, hoping to find work – like the men in the photo at the top of this page.

Populations of the cities named on the map * (millions)

Shanghai	15.6
Beijing	13.2
Guangzhou	11.1
Hong Kong	7.0
Tianjin	5.2
Chongqing	5.1
Wuhan	4.9
Harbin	4.8
Shenyang	4.5

* Does not include built-up areas outside the city limits.

How the country is divided up

Look at the map on the right. It shows how China is divided up. Henan is the most populous province, with about 99 million people. (The UK has 61 million.) Tibet is the least populous, with about 2.8 million.

Look at Hong Kong. It was a British territory for over 130 years, but was handed back to China in 1997. Macau was controlled by Portugal for centuries, but was returned to China in 1999.

China promised to let both places govern themselves, for 50 years from handover. That's why they are called 'special' regions.

Now look at the map on page 138. It gives the names of all the regions.

China's administrative regions

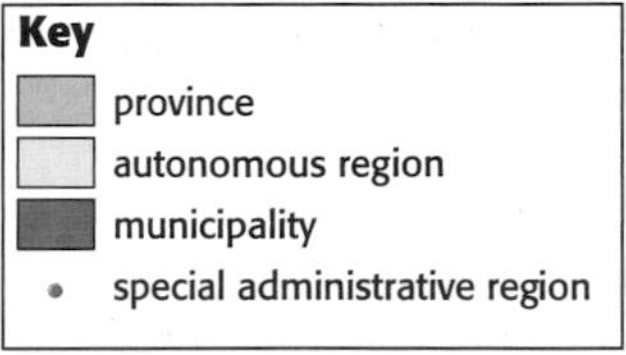

More about the people

China has 56 ethnic groups.

- Around 92% of the population belongs to the main group, the Han.
- The 55 minority groups include the Sani of southern China, and the Mongolians of northern China, and the Tibetans of Tibet.

Most ethnic groups live in the autonomous regions, where they can follow their traditional way of life, up to a point. Look at the map above. (*Autonomous* means you have some independence.)

Tibet is one of the autonomous regions. It was a separate country until 1950, when China invaded it and took over. The Dalai Lama, its spiritual leader, lives in exile in India. (Tibetans are Bhuddists.)

▲ *Women of the Sani ethnic group.*

Languages

China has seven languages, each with many dialects. The main language is Mandarin. The Mandarin dialect of Beijing is China's official language. All pupils must learn it at school.

Your turn

1 Look at the map on page 34. The population density around **P** is 1–10 people per square km.
  a What does *population density* mean?
  b What is the population density around: **i Q**? **ii R**?

2 Look again at the map. Which parts of China are:
  a the most heavily populated?
  b the least heavily populated?
  (Use terms like *west*, *east*, *coast* and so on.)

3 Now compare the map with those in Unit 2.1. Can you find any link between the pattern of population density, and:
  a relief? **b** the rivers shown on the map?
  c climate regions? **d** rainfall?
  If your answer to any of these is *yes*, see if you can *describe* the link, and then *explain* it.

4 Look at the cities named on the map, on page 34.
  a Which one is China's capital city?
  b Which is China's largest city?
  c Which might you expect to have sea ports?
  d Which are on the River Yangtze?

5 Now look at the map at the top of this page.
  a Do you notice any pattern in where the autonomous regions are located?
  b What can you say about their population density?
  c Why are Hong Kong and Macau 'special'?

6 China has a very large population – the world's largest.
  a About how many times larger is it than the UK's population? (The UK has 61 million.)
  b i Can you think of any difficulties a large population could cause, for a country's government?
  ii Can you think of any benefits it may bring?

2.4 China's one-child policy

Here you'll learn why so many Chinese children have no brothers or sisters!

A population crisis

For years, China's leaders were alarmed at how fast the population was growing. They said it was holding back development. So in 1979 they came up with the **one-child policy**.

The one-child policy

- Couples in towns and cities can have only one child.
- In rural areas, you may get permission to have two children, if both parents are only children, or the first child is a girl.
- If you have more children than your quota, you will be fined, and lose other benefits too.
- Ethnic minorities can have more than two children per family, since their numbers are low.

▲ *A poster for the one-child policy.*

Two people's stories

Wei, an only child

Hi, I'm Yang Wei. Yang is my surname. (We put this first, in China.)

I'm 14, and at middle school here in Shanghai. I have no brothers or sisters. In fact everyone in my class is an only child. My mum says she'd have liked more children, but we must put China first.

So, what's it like for us, being only children? It's good in some ways. Our parents do everything for us. They try to give us everything we want. If I had brothers and sisters I would not get so much attention.

But it's bad in other ways. When your parents have only you to think about, they put too much pressure on you, and especially about school.

I have to study all the time. And not just pass exams, but do really well. I got 76% in a maths test last term and they stopped my pocket money for a whole month! All they want is for me to get into a good high school, and then university, and get a good job.

It's the same for everyone in the class. Some of us study for 14 hours a day, between class and homework. I even work in the morning, before school. And my parents hate me to hang around with people who they think don't work hard.

There could be problems in the future too. When I get married, my wife and I will have to look after our child. And also help our four parents. And maybe even our grandparents, because they get only small pensions. It could be a big burden. So we'll need jobs that pay well.

Still, the policy may have changed by then. Or I might start a business and get rich. Then I might have more than one child – because fines would not bother me.

Did you know?

◆ A survey showed that over 75% of Chinese are in favour of the one-child policy.

What if ...

◆ ... nobody in your class had a brother or sister?

▲ *At school in Shanghai. Familiar?*

Ju, a farmer

I'm Tong Ju. My husband and I are rice farmers here in Yunnan province.

We already had two daughters when the One-Child Policy came along. Of course we love them. But here in China people think it is better to have a son. Because when a daughter marries, she can't really help you. You lose her to her husband's family.

Then I found I was pregnant again. The village committee called me in and gave me a hard time. And then our son Jian was born. We were so happy. We had to pay a big fine, but it was worth it.

Now our daughters are married in the next village. But our son went off to Ghuangzhou, because people can earn more in the city.

It can be hard for country people in the city, because they do not have a resident's permit. But Jian is smart, and works hard. He got a temporary permit, and has a good job in a factory. They made him a supervisor.

Jian got married two years ago, to Ting. Now they have a son. They work long hours, so they brought us the baby to look after. Jian sends money every month, and they come to visit for a week at Chinese New Year.

I do worry about the future. We don't have much money saved. There are no old-age pensions in this district yet. We do not want to be a burden on our son. Still, as my husband says, worrying never mended anything.

▲ *In China, grandparents play a big part in looking after children.*

Is it a success?

Yes, the one-child policy has slowed the rise in population. But it has brought problems too.

- Millions of girl babies have been abandoned, or aborted, so that people could try for boys. In 2007, 120 boys were born for every 100 girls.
- So in the future, millions of men won't be able to find wives.
- Also, in the future, young people may have to support several older people, and this could be difficult.

So the policy is being relaxed a little. Now it's easier to get permission to have two children in urban areas, if both parents are only children.

China's population

Year	Population (millions)
1910	410
1950	552
1965	725
1975	920
1980	987
1985	1059
1990	1140
1995	1211
2000	1267
2008	1320

Your turn

1 Wei's parents are keen for him to do well at school. Why is there so much pressure about this, in China?

2 Are China's leaders right to be concerned about population growth? You can draw a graph to find out! The population data is in the table above.
  - a First, draw and label the axes, with *Year* on the *x* axis.
  - b Plot the points, then join them with a smooth curve. Label your graph.
  - c The population in 1950 was 552 million.
    - i By which year had it doubled? (Use your graph!)
    - ii How many years had it taken, to double?

3 Look again at your graph. Can you see any evidence that the one-child policy is working? Explain.

4 a Now draw a dashed line on your graph to show how China's population might have grown since 1979, *without* the one-child policy.
  - b How did you decide where to draw your line?

5 Might China be *less* developed now, if it did not have the one-child policy? Think carefully about this. Then give your reasons.

6 The one-child policy has been stricter in urban areas. See if you can think of a reason for this.

2.5 How China is changing

The pattern of development around the world changes over time. China is a good example. You can find out more here.

> **Did you know?**
> ◆ By 2025, China could be the world's largest economy (in place of the USA).

A wealthy past

For centuries, China was the world's most developed and wealthy country. Even 200 years ago, it had the world's largest economy.

Much of this was due to its technology. For example, it was ahead in building canals, to move crops and goods around. It also had a civil service to help run the country. The officials were chosen by exam, and only the very best got through.

Trade with other countries was another big factor. Europe was mad for Chinese silk, tea, and porcelain. China demanded payment in silver.

▲ *As China develops, homes like this one are replaced by new apartments.*

China falls behind

From the **Industrial Revolution** (around 1750 onwards), countries in Western Europe developed rapidly. The UK became 'the workshop of the world'. Goods from its factories were sold everywhere.

But the Industrial Revolution did not spread to China. Instead, China suffered invasion, and conflict. So it fell behind.

In the 20th century, under Mao, China did develop some industry. But for much of that time it was shut off from the world.

Catching up fast

Now China is developing fast. This change is driven by **manufacturing** and **trade with the rest of the world**.

- Chinese people are now free to own businesses, and export goods, and make money for themselves.
- Thousands of new factories have been set up.
- A high % of the clothing, household goods, toys, and electronics on sale in our shops are made in China.
- In 2007, China became the world's top exporter of goods. (Germany was top before that.)
- China can sell goods more cheaply than most countries because it has lower wages.
- It also has a huge workforce. It has over 80 million workers in manufacturing. (The UK has 3.5 million.)

So China is having an industrial revolution now! It is often called a **newly industrialised country** (or NIC).

▲ *Something here for you? Containers of goods at the docks in Hong Kong, ready for shipping around the world.*

A rising standard of living …

Already, China's new industry, and trade with the world, has lifted hundreds of millions of people out of poverty.

There is more work. People are earning more. Their standard of living is rising. And the government has more to spend on schools and hospitals, and roads and other infrastructure.

Employment in China

sector	% of workforce
primary	43%
secondary	25%
tertiary	32%

▲ This shows China's main industrial areas.

▲ Chongqing, growing fast. On which river?

… but rising inequality

Most Chinese have benefited at least a little, from China's development. But some have gained a lot more than others. Overall:

- people in urban areas are much better off than people in rural areas
- the eastern half of China is much better off than the western half, because it has the industry. (Look at the map above.)

In fact the inequality between urban and rural people, and eastern and western China, keeps growing.

The government is working on these problems. For example it is helping Chongqing to grow fast, as a gateway to the west. (Look at the map.) It is building more roads in the west. And it plans to exploit the west's rich natural resources (like coal, oil, and gas).

GDP per capita for China

Year	GDP per capita (US dollars PPP)
2000	$4000
2001	$4020
2002	$4580
2003	$5000
2004	$5900
2005	$6800
2006	$7700
2007	$8900

Your turn

1 500 years ago, China was the world's wealthiest country.
- **a** The panel at the top of page 29 shows some Chinese inventions. Choose one that you think may have helped China develop before Europe. Explain why you chose it.
- **b** Now give a reason why China fell behind in development. (Unit 2.2 may help.)

2 a China is called a *newly industrialised country* (or NIC). What does the term in italics mean?
- **b** The UK is not an NIC. Why not?

3 Look at the map at the top of this page. Using the map to help you, see if you can explain why:
- **a** the west of China is less well off than the east
- **b** there is very heavy river traffic (boats and ships) on the River Yangtze

4 Look again at the map at the top of this page.
- **a** Note where the main industrial areas are located. See if you can find a pattern. (Hint: H_2O?)
- **b** Write a set of bullet points to *describe* the pattern.
- **c** Now see if you can *explain* the pattern.

5 Look at the table for GDP per capita, above.
- **a** First, what is: **i** GDP? **ii** GDP per capita?
- **b** How do you think the Chinese government felt about the *trend* in the table? Explain.
- **c** What is the reason for this trend?

6 As you saw, China has been developing fast. So how developed is it compared with other countries? This table will help you find out.
- **a** First, what is the HDI? (Glossary?)
- **b** HDI is better than GDP per capita as an indicator of development. See if you can explain why.
- **c** Now list the countries in the table, in order of development.
- **d** Comment on the order.

HDI values for a recent year

Brazil	0.807
China	0.762
India	0.609
Russia	0.806
UK	0.942
USA	0.950

7 So, China depends heavily on selling goods to other countries. But that can be risky too.
- **a** What does *recession* mean? (Glossary?)
- **b** See if you can explain this headline from 2009!

Recession around the world hurts China

Welcome to Beijing

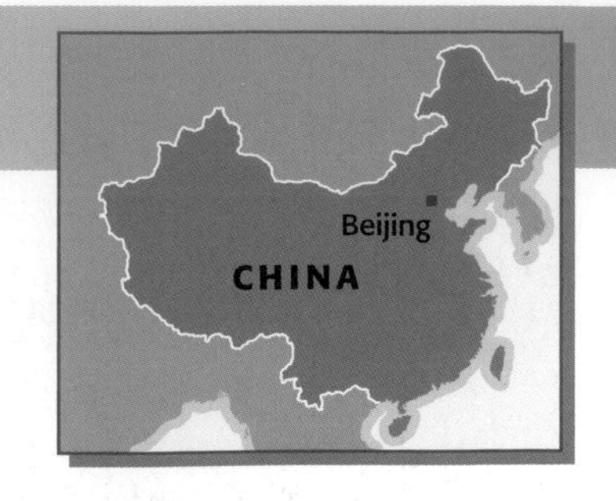

This unit is about Beijing, China's capital city.

Introducing Beijing

Beijing is a big bustling city of over 13 million people. It is one of the world's great historic cities. For centuries, it was home to the Emperors, who lived in the Forbidden City. Beijing grew outwards from there.

It has changed a lot in recent years. Many run-down areas have been swept away. There are new buildings, businesses, roads, airport terminal, and subways. Many changes were to get ready for the 2008 Olympics.

The shape of Beijing

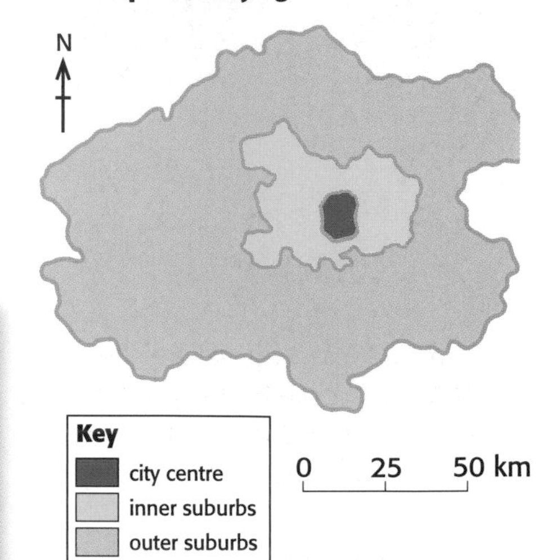

So what's it like?

Here's one person's description.

Hi Dan

Hello from the Peace and Harmony Hotel! Peace inside – but the little street outside is buzzing. It's a warm evening and the locals are sitting out on doorsteps and stools, chatting. From here I can see the lanterns at the noodle stalls, and just across the way, four old blokes are playing mahjong.

It's an amazing place. So many wonderful historic old buildings. And then the stunning new ones, like the Bird's Nest stadium. Hundreds of skyscrapers. Companies from all over the world. New blocks of high-rise apartments. Luxury homes for the new millionaires. And still there are the nice little streets like this one – a bit shabby, but full of life.

And then there's the shopping. From top designer shops to packed markets, selling everything under the sun. Even a tea market. Imagine hundreds of stalls selling tea! I like the late night food streets best, where you can sit and eat outdoors.

On Sunday we took a bus ride out through the suburbs. We saw lots of factories. We saw poor areas too. And farmland growing vegetables for the city. In the centre, it's easy to forget that China is still a developing country. Some of Beijing is far more modern than back home.

We walk a lot. But you take your life in your hands when you cross the street. Crazy drivers! And there is heavy smog some days, thanks to the non-stop traffic jams. Not nice for the lungs.

Tomorrow we take a taxi out of town, to visit the Great Wall. Then on Thursday we go south, by train. So e-mail us now!

Love from Sarah

▲ *That's better! Apartments in Beijing.*

Some faces of Beijing

The Forbidden City, where the Emperors lived. Ordinary folk were not allowed in. Now thousands of tourists stroll through every day.

The CBD and its new skyscrapers. Here you'll find the headquarters of banks, the media, and foreign companies working in China.

For many people, daily life is lived along bustling streets like this one. Rickshaws are a popular way for tourists to get around.

A thirsty city

Air pollution is a problem in Beijing, with all its cars and traffic jams. So are the dust storms that blow in from the west in spring.

There's an even bigger problem: water. The growing city needs more and more. This a dry area. The reservoirs that feed the city run lower every year. So the water supply is getting less and less reliable.

But Southern China has plenty of water. So the plan is to pump water from the Yangtze and its tributaries to feed Beijing and the dry north. It's called the **South-to-North Diversion project**.

The water will follow the three routes on this map. It will flow in existing river beds, and through new tunnels and canals. Work has started on the central route. The project should finish around 2050.

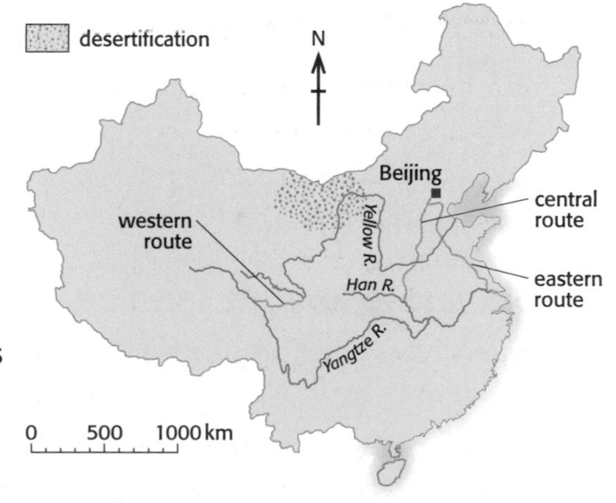

▲ *The South-to-North Diversion project.*

Your turn

1 Turn to the map on page 34, and see if you can give four facts about the *location* of Beijing.

2 Look at the photos in this unit. Choose the one you like best. Study it. Then write two paragraphs describing carefully what it shows, and why you selected it.

3 Now look at Sarah's description of Beijing, on page 40. See how many *geographical facts* you can pick out from it, that tell you about the geography of Beijing.

4 See if you can pick out any changes in Beijing:
  a that the Emperors would not have approved of
  b that Mao Zedong might not have approved of.

5 Look at the map of the South-to-North Diversion project.
  a Give a reason why the north of China has a water shortage, while the south does not. (Page 31?)
  b **i** About how long is the central route, up to Beijing?
  ii How does that compare with the distance from London to Edinburgh (530 km)?
  c If this project fails, what do you think might happen to Beijing?

6 Perhaps China should have chosen a different location for its capital city. Using the maps on pages 30, 31 and 34, see if you can suggest a better location. Give reasons for your choice.

2.7 Life in rural China

What's life like, for China's farmers? You can find out here.

The rural population

About 730 million people live in rural China – or 56% of the population.

Most live by farming. The farmers lease plots of land to farm. (They do not own the land.) They usually have to sell some produce to the state. They can sell the rest in markets.

Some crops grown in China		
grains	**other food crops**	**non-food crops**
rice wheat corn	soybeans sugarcane tea peanuts	tobacco cotton mulberry (for silkworms)

A tough life

Life has improved for most Chinese people, over the last 20 years. But people in urban areas have gained most. On average, farmers earn about a third of what people in urban areas do. Life is tough in other ways too.

- There are some large farms, using modern machinery. But most are small, and the farmers have to do most of the work by hand. (Many can't afford machinery.)
- The state does not pay that much for the food it buys, because it wants to keep food prices down for city dwellers.
- Many rural schools are poor. You don't learn much. So when people move to the cities they can get only low-paid jobs.
- Health care is poor too. And many can't afford to go to the doctor.

There are lots of happy farmers, who love farming, and do well. But around 100 million Chinese still live in poverty – and most are in rural areas. The further inland you go, away from the coast and big cities, the more poverty you'll find.

▲ *A rural village in China.*

You can lose your land

Farmers have other problems too. Many have been thrown off the land, to make way for factories and other developments.

Since the farmers don't own the land, they often don't have much say. Decisions are made by local officials, who can sell land for high prices. But the farmers often get little compensation. They think this is unfair. There have been thousands of protests.

▲ *A water buffalo helps with ploughing, in a rice field.*

▲ *Lunch break. They left the farm for building work in Beijing.*

So, head for the city?

What can a poor rural family do, to earn more? Usually, someone goes off to a city to find work. Like building work, or factory work, or a job as a street cleaner or market porter.

The **migrant workers** don't earn much per hour. But they work hard to send money home. Looking after the family is a duty.

In the cities, they are treated a bit like illegal immigrants, because they don't have residents' permits. In the past, they were often arrested and sent home again. But the building sites and factories need workers, so the rules were relaxed.

▲ *Some homes in Huaxi. It is called China's richest village.*

Or urbanise?

Some rural villages have taken a bigger decision: stop farming, and build factories instead! That's what they did in Huaxi, near Shanghai.

Once a poor and sleepy village of 1500 people, Huaxi is now home to nearly 60 companies, and over 30 000 people. The village families still own the land collectively, and share income from the factories. Now they are all rich!

A big dilemma

China has one big dilemma: not enough farmland. It has 20% of the world's population – and less than 7% of its farmland. And it has lost a lot of farmland to **urbanisation** and other uses, over the last 25 years.

- So the government is worried about food shortages in the future.
- It has set a 'bottom line' for the amount of land to be kept for farming. (At least 120 million hectares.)
- It plans to modernize farming, fast. So farmers can now lease or sell their 'rights' to plots, to other farmers. The aim is to have big farms, with modern machinery, producing as much food as possible.
- The government also plans to buy or lease farmland in other countries with unused land. For example countries in Africa and South America. (It has leased quite a lot already.)

The government has promised to raise farmers' income too, and improve rural schools, and health care, and roads and other infrastructure.

▲ *Thousands of rural Chinese have moved to Africa, to work on Chinese projects or set up on their own.*

Your turn

1 See if you can give reasons why China's government:
  a wants to keep the farmers happy
  b would prefer larger, more modern farms
  c would like to buy land in other countries

2 a Find the term in the text that means *people who move around to find work.*
  b In Mao's time, people from rural area were not allowed to move to urban areas. If this were still the case, how do you think it would affect:
    i rural areas? ii urban areas?

3 a What does *urbanisation* mean? See if you can work it out, then check in the glossary.
  b Now give an example of urbanisation, from this unit.

4 You are the Prime Minister of Ghana. The Chinese government wants to lease some good farmland in Ghana. It will send farm machinery, and experts to help improve poor land in other parts of Ghana.
  a Give a list of reasons for saying *Yes* to this idea.
  b Now give a list of reasons for saying *No.*
  c What will your final decision be?

2.8 What about the environment?

Here you'll learn about the challenges China faces, in cleaning up its environment.

Develop first, clean up later

In Britain, development took off with the Industrial Revolution.
We set up lots of factories. Towns and cities grew rapidly. We poured harmful gases into the air, and toxic waste into rivers. We ruined places.

And then, when we got richer, we began to clean up.
Most other rich countries have followed the same pattern.

▲ *Air pollution under way. At this plant in China, they are baking coal in coal-fired ovens, to get coke to make steel.*

So how is China doing?

China is developing faster than any country in history. It is making the same mistakes as Britain did. And now pollution is a massive problem.

Coal: the main culprit

China has lots of coal. It depends on coal for 70 % of its energy needs. Coal is burned in most of its power stations, and in factory furnaces.

So China's development is fuelled by coal, just as Britain's was. And the trouble is: coal is a very dirty fuel.

First, it gives off carbon dioxide when it burns. This is linked to **global warming**. Acidic gases (sulphur dioxide and nitrogen oxides) form too. They cause **acid rain**. Then particles of soot and ash get everywhere, including onto your washing, and into your lungs.

Other causes of pollution

- Factories pollute the air and rivers with toxic chemicals

  There are laws against this. But often they are not enforced – because producing lots of goods is the priority.

- More and more people have cars. So carbon dioxide and other gases from car exhausts add to the problem.

- There are not nearly enough sewage works. So millions of tonnes of untreated waste from homes, and factories and other businesses, pour into streams, lakes, and rivers every day.

A small shop beside a coal mine. Hard to keep everything clean? ▶

The clean-up begins

China is now very worried about pollution. It is also under pressure from other countries, because it is the world's top producer of carbon dioxide. So it is trying to clean up.

- It plans to build hundreds more coal-fired power stations. But scientists are looking for ways to trap the carbon dioxide from them.
- It plans to build more nuclear power stations, and wind farms, and dams for hydroelectricity, since these don't give carbon dioxide.
- It has started to build thousands more sewage plants. (It needs at least 150 000 more.)
- It is being tougher on factories that cause pollution.

▲ *One of China's windfarms.*

Another problem: desertification

In northern China, there's another problem too: farmland is turning into desert. The Gobi desert is spreading. (Look at the map on page 30.)

Why? Because, in the farmland around the Gobi:

- herders let their animals overgraze; so every scrap of vegetation gets eaten away.
- trees have been chopped down.
- farmers have farmed the soil too intensively, and taken too much water from rivers and wells to water the crops. So the water table has fallen. Rivers are drying up. The soil is drying out and growing useless.

In spring, the wind carries the dry exposed soil eastwards. The dust storms choke Beijing – and even reach Japan.

So now the Chinese are planting a belt of trees and shrubs, to try to stop the advance of the desert. They call it the Green Great Wall. When finished (by 2075) it will be 4500 km long.

Work has started too on an even more ambitious project. They plan to pump water up from the Yangtze, to water the dry north. There is more about that on page 41.

▲ *A farmer walks home in a sand storm in Gansu province. (Look at the map on page 138.)*

Your turn

1 Say what these terms mean. (Try without the glossary?)
  a pollution **b** acid rain
  c global warming **c** desertification

2 The table below lists environmental problems in China.
  a Make a *large* copy of the table. (Use a full page.) Give your table a title.
  b Try to give two causes for each problem, from this unit.
  c Then see if you can give at least one consequence for each. Use your own store of knowledge too.

Environmental problems	*Causes*	*Consequences*
air pollution		
water pollution		
desertification		

3 The first section at the top of this page shows steps China is taking, to tackle pollution.
  a Which do *you* think is the most important step it is taking? Explain your choice.
  b Would these choose the same as you? Give reasons.
    i the man in the photo at the top of page 44
    ii the panda on page 28 (if it could …)

4 What would you say, in reply to these opinions?

a

b

2.9 The Three Gorges dam

In this unit you'll learn about China's Three Gorges dam: the largest dam in the world.

▲ *The Three Gorges area.*

More electricity please!

China has over 1.3 billion people. It is developing fast. New factories and other businesses have sprung up. People are earning more – and buying more and more appliances for their homes. So China needs more and more electricity. It already has shortages!

Hydro

80% of China's electricity is from burning fossil fuels – mainly coal.

Most of the rest is hydroelectricity. China has many fast flowing rivers, thanks to the steep land in the west. It has *thousands* of dams of all sizes, from mini to large, generating electricity.

The new Three Gorges Dam is the largest. It's the largest dam in the world.

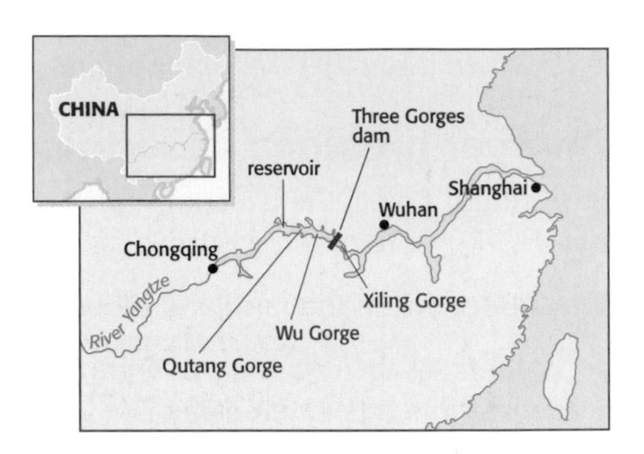

The Three Gorges dam

The dam is built across the River Yangtze, in a steep valley.

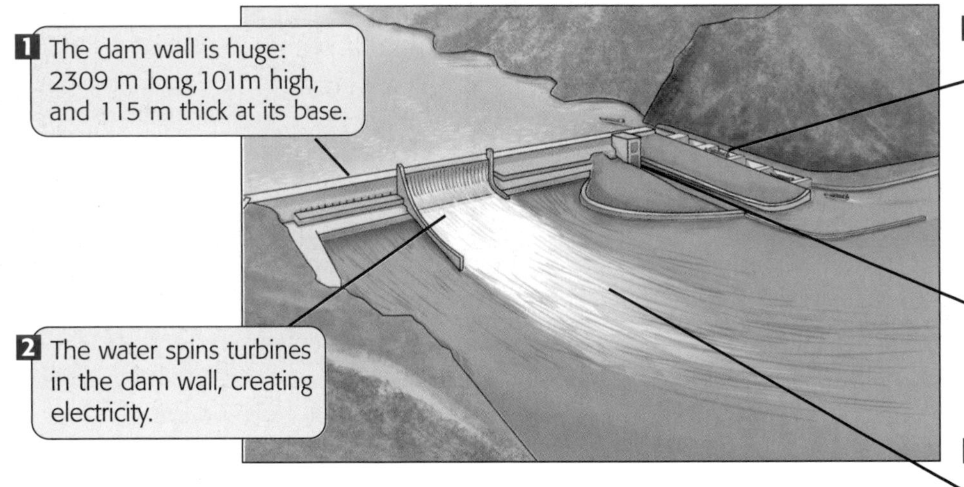

1 The dam wall is huge: 2309 m long, 101 m high, and 115 m thick at its base.

2 The water spins turbines in the dam wall, creating electricity.

3 Ships can climb up and down beside the dam using a system of locks (water basins).

4 A massive lift is also being built, to raise and lower large ships.

5 The dam is already generating electricity. When it is complete in 2012, it will provide over 2% of the electricity China needs.

Is it a sustainable development?

The Three Gorges dam is an amazing feat. But is it a good example of **sustainable development**? Study the next page, and see what you think.

Your turn

1 What is a *dam*?

2 See if you can give a reason for selecting:
  a the Yangtze b the 'Three Gorges' area
  as good places for a dam. (Units 2.1, 2.3, and 2.3?)

3 China hopes the dam will improve the standard of living, for its people. Give three ways it could do this.

4 Sustainable development brings *economic*, *social*, and *environmental* benefits. Explain the terms in italics.

5 a Make a much larger copy of the table started below, with plenty of room to write in the spaces.
  b Now fill in as many benefits and negative points as you can, in the correct places, for the dam.

6 So is the dam a good example of sustainable development? Answer in *not less than* 30 words.

The Three Gorges dam	Economic	Social	Environmental
benefits			
negative points			

A The economy

- The dam will have cost about £27 billion by the time it is finished.
- The electricity will benefit many thousands of factories, other businesses, and homes.
- Thanks to the deep reservoir behind the dam, large ships can now go further up the Yangtze than before, carrying cargo.
- This is great for towns and cities along the river. Industry is booming in Chongqing city!

B No more flooding?

Electricity was one reason for building the dam. Another big reason was flood control.

- Over the centuries, flooding on the Yangtze has killed hundreds of thousands. In 1954, floods drowned over 33 000 people, and left the city of Wuhan under water for over three months.
- The dam controls the flow of water. So it is hoped that future flooding can be avoided.
- Some say the dam is causing the opposite problem! At drier times of year, downstream from the dam, the river level is falling really low.

C Impact on local people

When you dam a river, a reservoir of water builds up behind the dam wall, drowning everything in its way.

- Around 1.25 million people were moved, to make way for the Three Gorges reservoir.
- They left behind 2 empty cities, 116 towns, and hundreds of villages. These were drowned forever.
- So were farms, family graves, and historic sites. Many people felt very unhappy about all this.
- People got some money as compensation, but they say it was not enough.
- Some people were given plots of land to move to. But many moved to Chongqing to find work. (Chongqing is growing rapidly.)

D The environment

- Hydroelectricity is clean energy – no fuel is burned, and no carbon dioxide is formed. So the dam is helping to fight global warming.
- But it stops fish and other river animals from moving freely. The rise in shipping affects them too. Now the Yangtze River dolphin has almost died out.
- Some people say poisonous levels of sewage and factory waste will build up in the reservoir. This would harm wildlife, and humans.

▲ A Yangtze River dolphin. Most are now dead, because of the dam. Experts say there are too few left to breed. So they will soon be extinct.

E New dangers?

Some say the new dam brings other dangers.

- The area is prone to earthquakes. And the weight of water in the reservoir puts extra strain on the surrounding rock.
- Filling the reservoir has already caused many small earthquakes, and landslides, near the river.
- If the dam cracks, the rush of water downstream will cause a catastrophe.
- In any case, the reservoir may fill up with silt, making the dam less and less efficient.

The people in charge of the dam know they can't relax. They must always watch out for danger.

China's place in the world

Here you'll find out about China's growing influence in the world.

A growing power

China is a big country. A fifth of all humans live there. It has opened its doors to the world, and is developing fast. So it is growing more and more important.

In fact China affects billions of people all around the world – including you! Find out more in these panels. (You have met some of the information already.)

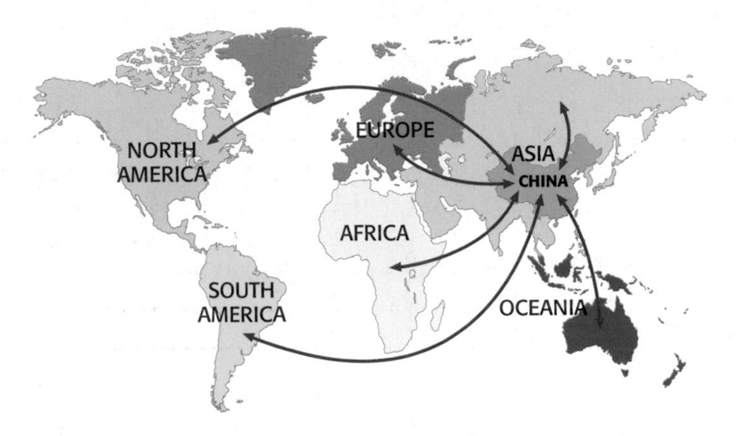

1 Chinese culture

- You have probably used some Chinese inventions. Like paper, a wheelbarrow, a compass, fireworks, kites …
- And some of these may have touched your life: kung fu, acupuncture, Chinese food …
- Many cities around the world have a 'Chinatown', where you can eat, shop, and visit herbal doctors.

▲ *London's Chinatown.*

2 Made in China

Some of your things were almost certainly made in China. It is often called 'the world's factory'.

- It is now the world's top exporter of goods.
- It exports everywhere, including the UK. But its main trading partner is the USA.
- It also buys lots of things from other countries. Especially raw materials, as you'll see in panel 4.
- Overall it earns much more than it spends. In 2007 it earned 262 billion dollars more! This is helping it to develop.

▲ *On the way …*

3 The flow of work to China

Many factory workers in richer countries have lost their jobs to China. (Including people in the UK.)

- That's because many big companies (TNCs) have closed their factories in richer countries – and opened in China instead.
- They go to China because wages are lower there.
- Other foreign companies get goods made in Chinese factories, for the same reason.

▲ *China is the world's top manufacturer of clothing.*

4 Buying up resources

As China develops, it needs more and more resources. It does not have nearly enough of its own.

- So it buys resources from other countries. Like oil, steel, other metals, and wheat.
- It is becoming the world's top buyer of resources.
- This causes world prices of some resources to rise, since other countries want them too.

▲ *The Bird's Nest stadium in Beijing: lots of steel.*

5 Investing in other countries

China is also investing in other countries – often to make sure of resources. For example:

- It has bought a share in many oil fields and mines around the world.
- It has bought or leased land to grow soya and other crops, oil palm trees, and trees for timber.
- It has set up factories in countries with even lower wages, making things to sell there.

▲ *China has invested in most countries in Africa.*

6 Aid to other countries

Many poorer countries receive aid from China.

- It is often given in return for permission to buy a share in an oil field or mine.
- It is often in the form of new roads, railways, and other infrastructure that the country badly needs.
- Chinese workers are sent to help build these.

▲ *A new road for Ethiopia in Africa, thanks to China.*

7 Politics

- China has great political influence.
- In particular, it is on the United Nations Security Council (with the UK, USA, France, and Russia).
- The Council's aim is to keep peace in the world, and settle conflict by negotiation.

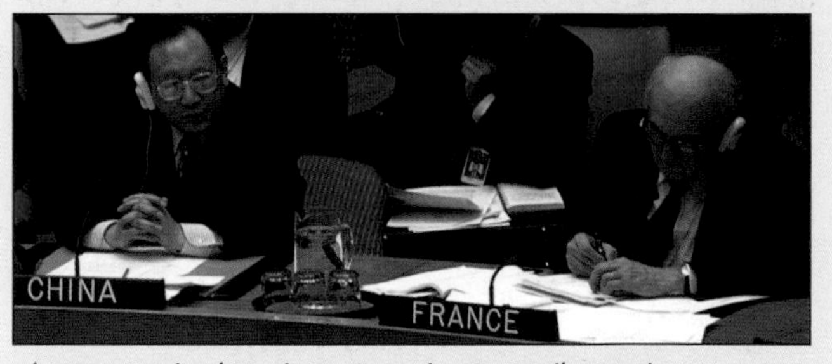

▲ *At a United Nations Security Council meeting.*

Your turn

1. How many links can you find, between China and *you*? Design a diagram to show them. Try for at least five.
2. See if you can explain why China might want to:
   - **a** buy a share in a big oil field in Sudan
   - **b** buy farmland in Zambia
   - **c** set up a bicycle factory in Ghana
3. Some people say China is colonising Africa, like Britain and other countries did in the past. Do you agree with that opinion? Explain.
4. Where does China get the money for all its investments in other countries?
5. Now design and draw a diagram to show what is flowing out of China, and what is flowing into it, to show China's importance in the world.
6. Pretend you are China's leader, for this.
   - **a** Panels 1–7 show China's links with the rest of the world. Which panel is the most important, to you?
   - **b** Put the panels in order, the most important first.

3 Off to the USA

Hop on a plane, head across the Atlantic, and around 7 hours later, you'll be in the USA! Where you will find …

▲ *… some world-famous landmarks …*

▲ *… world-famous companies …*

▲ *… stunning scenery …*

▲ *… great cities …*

▲ *… wealth and glamour …*

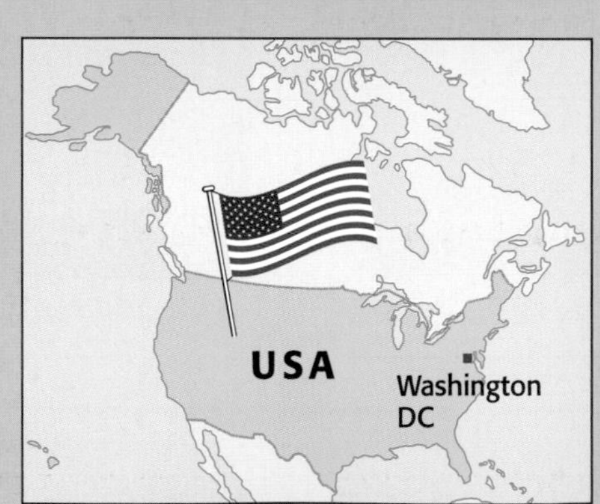

▲ *… poverty …*

▲ *… people of every nationality …*

▲ *… and a passion for baseball!*

The big picture

This chapter is about the USA. These are the big ideas behind the chapter:

- The USA one of the world's most developed and wealthy countries.
- It is made up of 50 states, each with its own state government.
- It has been built up by immigrants and their descendants, over 500 years.
- We think of it as the world's superpower.
- But like every country, it has problems to tackle.

Did you know?

- 4 July is Independence Day in the USA.
- On 4 July 1776, 13 British colonies in North America declared they were no longer British.

Your goals for this chapter

By the end of this chapter, you should be able to answer these questions:

- Which physical features of the USA can I name, and place on a map? (At least six, including at least two rivers.)
- What is the climate of the USA like?
- How many people live there? And what kind of racial / ethnic mix?
- Which parts of the USA are the most highly populated? And why? (See if you can give at least three reasons.)
- Which American cities can I name, and place on a blank map? (At least five.)
- What is the American Dream? And does it come true, for everyone?
- What can I say about:
  the car industry in the USA? farming in the USA?
  (See if you can give four facts about each.)
- How much can I say about the geography of California? (Give *at least* two facts about its climate, four about its physical features, and three about its human geography.)
- The USA is considered a superpower. Why?
- The USA has some serious problems to tackle. Some examples are … (Give at least three examples.)

Did you know?

- During the War of 1812 between Britain and the USA, the British Army set the White House on fire!

Did you know?

- The state of Alaska has over 3 million lakes!

And then ...

When you finish this chapter you can come back to this page and see if you have met your goals!

What if...

- ...the USA turned history upside down, and took Britain over?

Your chapter starter

Look at the photos of the USA, on page 50.

What does 'USA' stand for?

Do you recognise anyone, or anything, in the photos?

What other photos would *you* add, to represent the USA?

Does the USA have anything to do with you?

The USA: physical geography

Here you will find out about the main physical features of the USA, and its climate.

It's big!

The United States of America is the third largest country in the world, after Russia and Canada.

It's over 40 times bigger than the UK.

Look at **Alaska**. It is one of the 50 states that form the USA. But see how it's cut off from the rest.

And look at **Hawaii**, the group of little islands in the Pacific. It is one of the 50 states too.

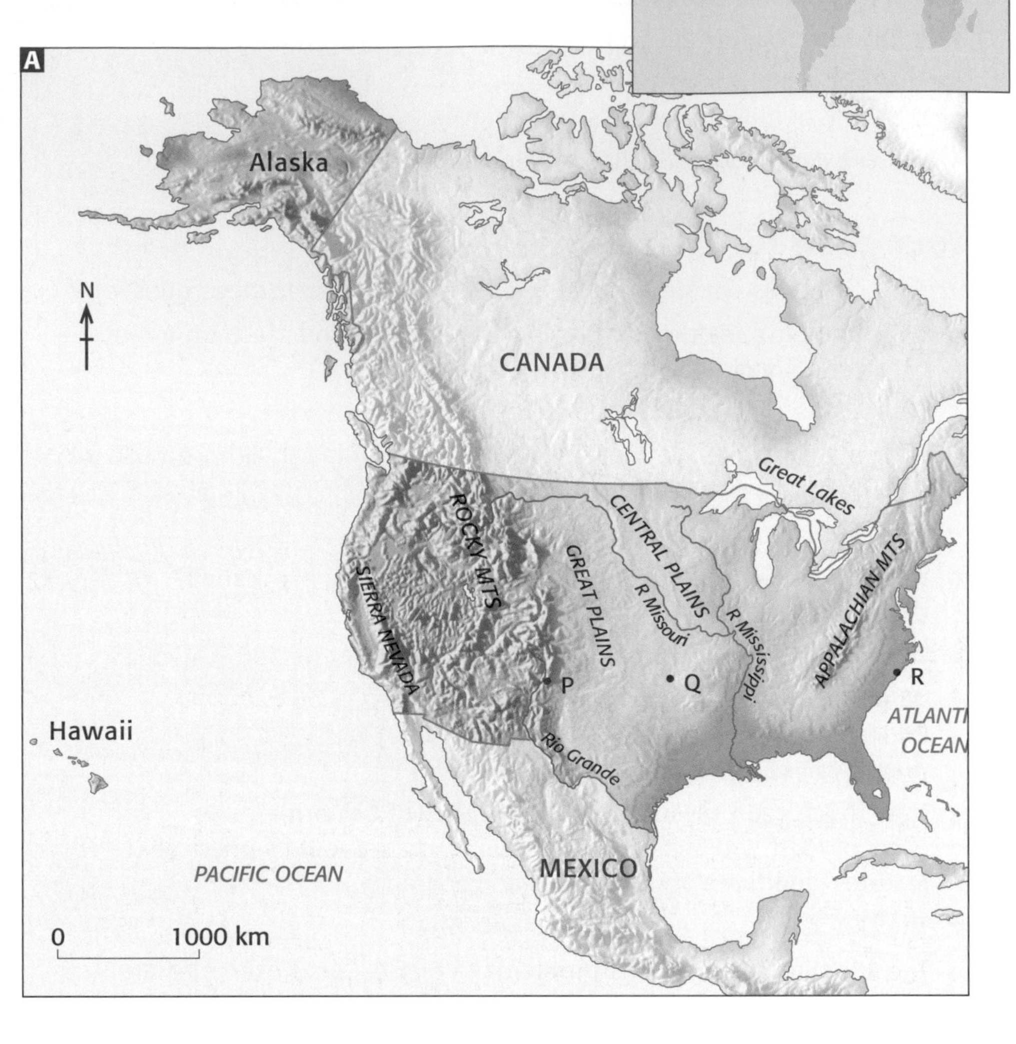

▲ *Mt McKinley in Alaska is the USA's highest peak (6194 m).*

Relief

- Look at the map. The eastern half of the USA is low and flat – apart from the **Appalachians**, which are low rounded highlands.
- The **Central Plains** are flat or gently rolling plains.
- The **Great Plains** are a little higher. With the Central Plains, they form a key farming region: for cattle, and crops like wheat, corn, and soybeans.
- Now look at the mountains in the west. They run from Alaska down to Mexico. The range of **Rocky Mountains** has the highest peaks.

Rivers and lakes

The USA has more than 250 000 rivers! The map shows just three.

- The Missouri is a tributary of the Mississippi. Together, they form a river nearly 7000 km long: the fourth longest river in the world.
- The Rio Grande, 3000 km long, forms part of the border with Mexico.
- Look at the huge Great Lakes. They are the largest group of lakes in the world. Four are shared with Canada.

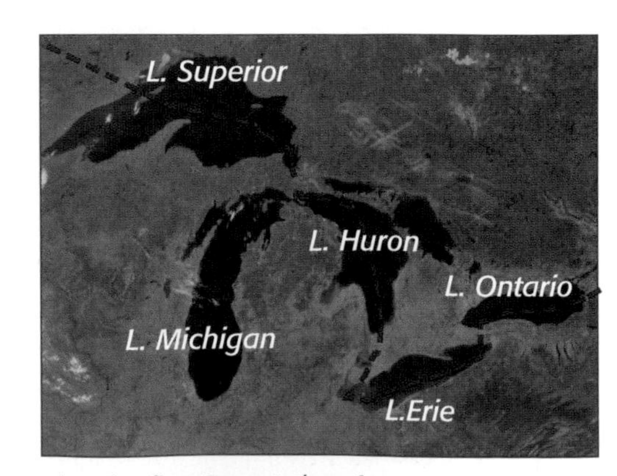

▲ *The five Great Lakes, from space. The red dashed line has been added to show the border with Canada.*

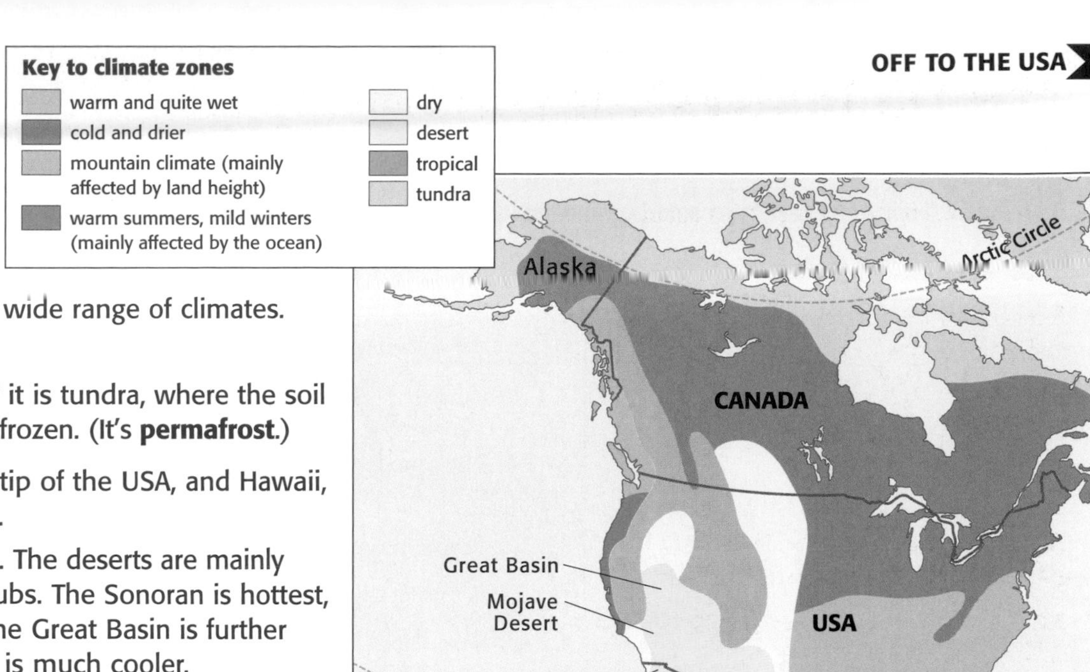

Climate

The USA is big, so it has a wide range of climates. Look at this climate map.

- Alaska is cold! Much of it is tundra, where the soil under the surface stays frozen. (It's **permafrost**.)
- But the most southern tip of the USA, and Hawaii, enjoy a tropical climate.
- Look at the desert area. The deserts are mainly stony soil, with low shrubs. The Sonoran is hottest, but has cool winters. The Great Basin is further north, and higher. So it is much cooler.
- Large land masses heat up fast in summer, and cool fast in winter. So inland, away from the sea, you find big temperature differences between summer and winter.

Natural hazards

- Earthquakes and eruptions are a hazard down the West Coast, because two of the Earth's **plates** meet there. Look at map **C**.
- Major hurricanes often strike the south east coast.
- Inland, powerful whirlwinds called **tornadoes** are quite common in spring and summer. They can carry houses away!
- Droughts occur in many parts of the USA. They are due to low rainfall, and often lead to **wildfires**. But they are not *completely* natural. Humans add to the problem, by pumping too much water from rivers and aquifers.

Experts say that droughts, and hurricanes, and tornadoes, are likely to become more frequent and severe, thanks to global warming.

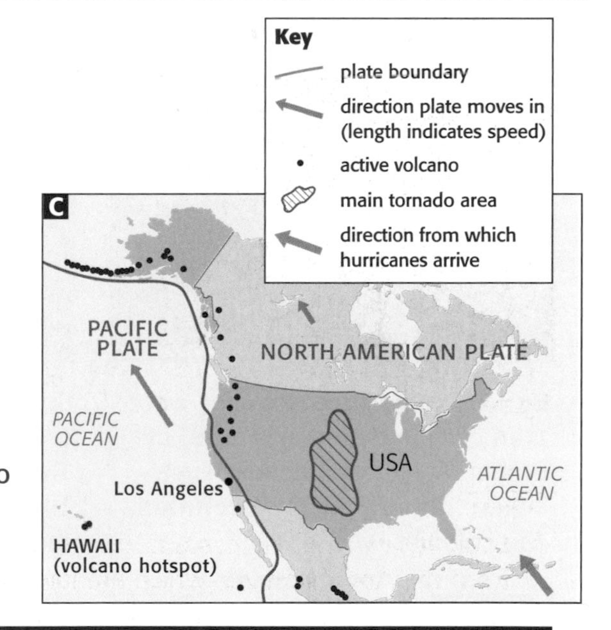

Your turn

1. **a** Which continent is the USA on?
   b Which countries border it?
   c Which oceans border it?
2. Now write a summary about the pattern of relief in the USA, as shown on map **A**. (See if you can do it *without* reading the text again. Just use the map.)
   Include these terms in your answer: *west, east, lowest, highest, coast.*
3. *The USA is a country of contrasting climates.* Do you agree with this statement? What is your evidence? (Map **B** may help.)
4. Using map **A**, see if you can explain *why*:
   a there is tundra in Alaska
   b Hawaii is much warmer than Alaska
   c it is always cooler at **P** than at **Q**
   d it is warmer at **Q** than at **R**, in summer
   e it is warmer at **R** than at **Q**, in winter.
5. Using map **C**, and what you know about *plates* from before, see if you can explain why:
   a the west coast of the USA has many volcanoes
   b Los Angeles is at high risk of earthquakes
   c there's a string of little islands off Alaska.

A little history

How did all those American states emerge? And get united? You can find out here.

The United States

People first reached North America over 12 000 years ago. They crossed from Asia to Alaska, on land that is now under water. Today's Native Americans are their descendants.

European explorers began to arrive around 1500 AD. And by 1733, Britain had 13 colonies along the east coast. France, Spain, and other countries took over other areas.

Some Europeans set up plantations. They needed workers, so bought hundreds of thousands of African slaves. The British sent convicts too. Poor Europeans came as servants.

By 1775, the British colonies were tired of being ruled and taxed by Britain. A war for independence began: **the American Revolution**. The colonies won.

So in 1783, Britain granted its 13 colonies their independence. They became **the United States**. They chose George Washington as their first president.

By 1900, 117 years later, the United States had reached its present size. It had spread by buying land, and gaining land through wars, conflicts, and treaties. (See page 139!)

The states did split up for a short time, from 1861. The president then, Abraham Lincoln, was against slavery. 11 southern states were in favour of it. They broke away ...

... to form their own **Confederacy** of States. This led to the **American Civil War** (1861 – 1865). The Confederates lost. The states were reunited, with slavery abolished.

Today, the 50 states are strongly united, under the American flag. It is called the Stars and Stripes – or sometimes the Star-Spangled Banner, after the national anthem.

▲ *This white building in Washington is the seat of government. It is called the Capitol. Slaves helped to build it.*

▲ *Barack Obama became the first black President of the USA, in January 2009.*

So who governs the USA?

The state governments

Each state has a capital city, and a state government. The state government looks after things like education, public health, roads, and law and order, for that state.

The Federal Government

Above them all is the **Federal Government**. It is based in Washington, DC, which is the USA's capital city. (DC stands for District of Columbia.)

The Federal Government is in charge of the defence of the USA, and its relations with other countries, and the overall USA budget. The laws it passes apply to all states.

The **President of the United States** is head of the Federal Government.

Note that Washington, DC is not like other American cities. It is not part of any state. It sits alone. Its main function is to be the seat of government.

▲ *The White House in Washington, where the president lives. Slaves helped to build it too.*

Your turn

1 See if you can explain why:
- a English is the main language of the United States
- b the Stars and Stripes has 50 stars
- c the Stars and Stripes has 13 stripes
- d almost 13% of Americans are black.

2 The map on page 139 shows the 50 states of the USA, plus the city of Washington, DC. See if you can name:
- a the largest state
- b the smallest state
- c the one furthest north
- d the one furthest south
- e three states that border: i the Pacific Ocean ii the Atlantic Ocean iii any of the Great Lakes

3 From the map on page 139, see if you can name:
- a three states on land bought from other countries
- b three on land gained since 1800, through wars, conflicts, and disputes with other countries
- c six that were part of the original 13 colonies (some have grown since then; some have split into two)
- d the last two areas to join the United States.

4 Now look at the USA's border with Canada. What do you notice about it? See if you can suggest a reason.

5 See if you can guess how the expansion of the United States affected the Native Americans.

A nation of immigrants

Here you'll find out about the population of the USA, and how it is spread.

Third largest for population

The USA has a population of 304 million. So it is the third largest country in the world, by population. (China is first, and India second.)

A nation of immigrants

Once upon a time, only Native Americans lived in the land that is now the USA. They had lived there for thousands of years.

Then, over 500 years ago, the first European explorers arrived. It was the start of a flood of arrivals.

- Many people came to seek their fortune: to get land to farm, or find other work.
- Some were brought by force. At least 640 000 African slaves were shipped in, between 1600 and 1865. Around 50 000 British convicts were sent over.
- Some came to escape persecution, or death. For example thousands of Irish people arrived to escape a famine in Ireland (1845 – 1850). Many Jews fled there from Europe, during World War II.
- In the last 50 years, many highly-skilled people have arrived from all over the world, including the UK, looking for better jobs.
- But overall, most immigrants have been ordinary people, without much education, ready to work hard and build a better life.

▲ *New arrivals from Italy. This photo was probably taken around 1900.*

▲ *Get those passports and visas ready!*

Still arriving

Immigrants are still arriving – around 1.6 million people a year.

They are from all over the world. But over half are from Mexico, and Central and South America. This group is called **Hispanics**.

It is thought that over 30% of the new immigrants are illegal.

The mix

Look at these pie charts.

A shows the mix of people living in the USA in 2007.

The 'whites' are mostly Americans of European descent, and more recent arrivals from Europe.

Most of the 'black' group are descended from slaves.

B shows new immigrants in 2007.

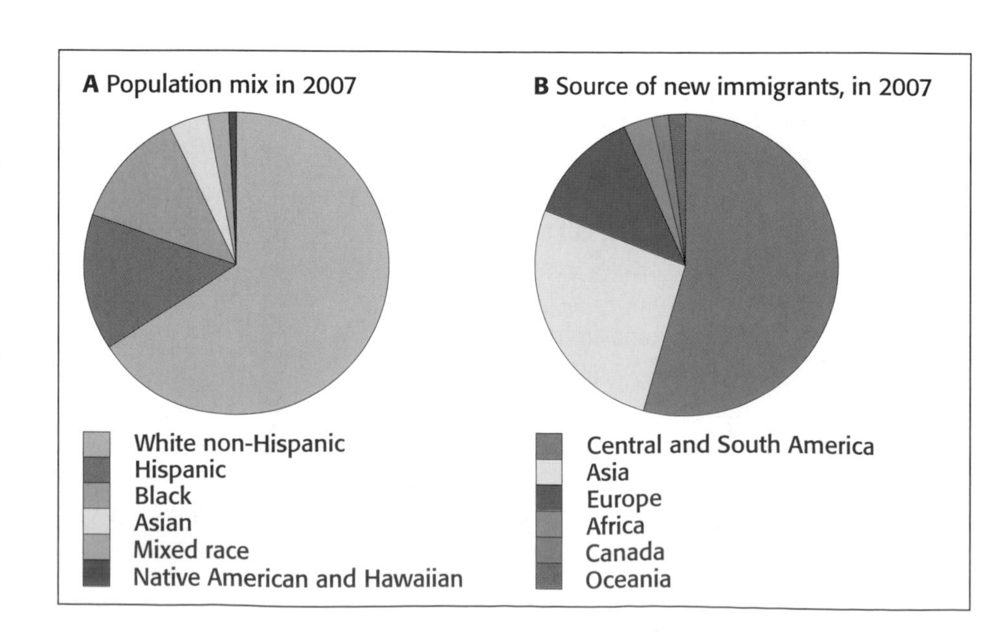

How it may change

Today, almost 70% of the USA's population is white.

But experts predict that by 2050, less than half will be white.

So where is everyone?

This map shows how the population is spread around the USA.
The tiniest white dot represents 7500 people.
So the larger the white patch, the more people in that area.

The USA's five largest cities

City	Population (millions)
New York	0.1
Los Angeles	3.8
Chicago	2.8
Houston	2.0
Philadelphia	1.5

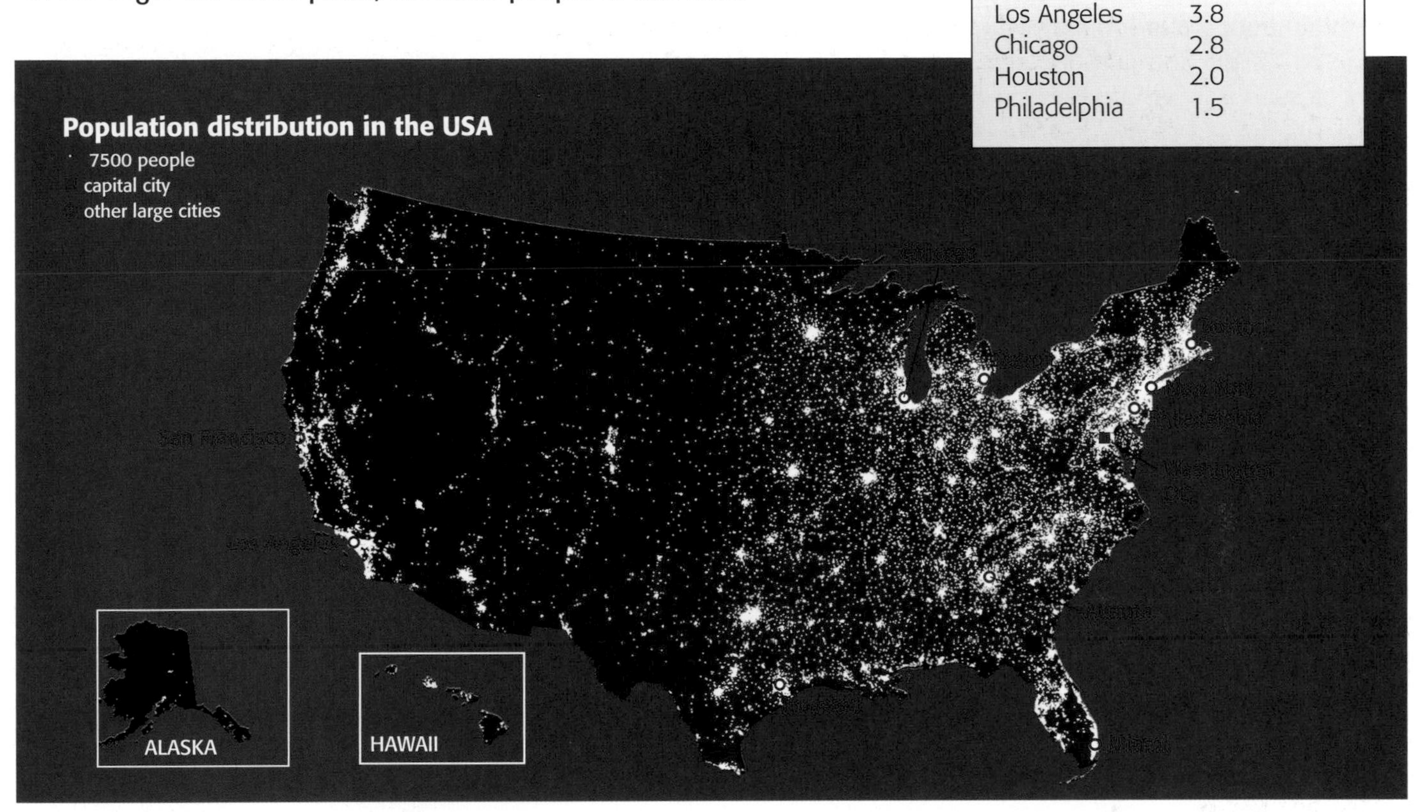

Your turn

1 First, write down what these terms mean. (Glossary?)
  a push factors **b** pull factors

2 People have been emigrating to the USA for hundreds of years, and still are, today.
  a Make a table with headings like this:

Reasons people have moved to the USA	
push factors	**pull factors**

  b Now see how many push and pull factors you can fill in. Page 56 may help, but think up others too.
  c Underline any factors that might lead *you* to the USA!

3 Using the pie charts on page 56 to help you, copy and complete this paragraph.
In the USA today, about 66% of the population is _______. The next two largest groups are _______ (14%) and _______ (13%). The smallest group is the _______, who make up less than 1% of the population. Over half of new immigrants these days are _______. They come from _______.

4 **a** Now make a *large* copy of the table started below. (You could turn your page sideways.)

Population distribution in the USA	
The patterns	**Possible reasons**
1 Overall, the most densely populated areas are on the coast. 2 The east coast is more densely populated than the west coast. 3 The eastern half of the country is more densely populated than the western half. 4 Alaska is 5	

  b Complete statement 4 in your table.
  c See how many more statements you can add. (For example, what's the population density like around the Great Lakes?)
  d Now, in the second column, see if you can explain the patterns. Maps in Unit 3.1 may help.

3.4 The Native Americans

This is about the people living in North America long before the Europeans arrived.

The first arrivals

We think groups of people arrived in North America over a long period, between 12 000 and 40 000 years ago. They were hunters from Asia. They crossed from Asia on a strip of land that is now under water.

These groups spread through North America, and into South America. Some developed great civilizations. (For example, the Inca and Aztecs.)

Today, their descendants include the Native Americans of the USA, and the rainforest tribes of the Amazon in South America.

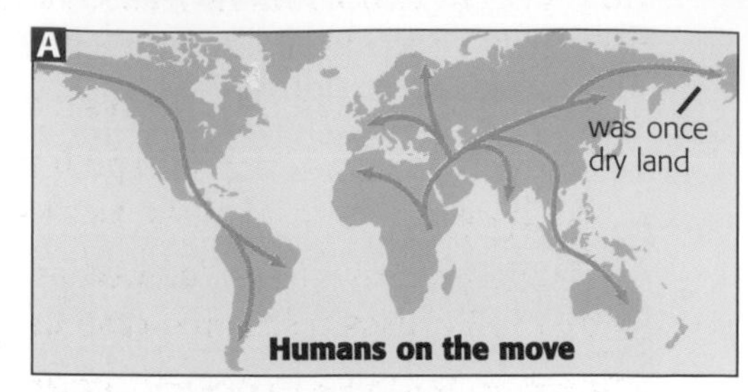

▲ *We evolved in East Africa over 200 000 years ago, and slowly spread.*

Where the tribes lived

There were hundreds of different tribes, living all over the continent. Map **B** names some. The colours show regions where tribes shared a similar way of life.

Adapt and survive

The tribes adapted to their environments, to survive. For example:

- The Inuit of the freezing Arctic hunted seal, walrus, bear, whale, and reindeer (caribou), for food. They burned whale oil, and dressed in animal skins.
- The tribes of the Plains lived by hunting buffalo. They used every part of the buffalo: flesh for meat, skin for clothes and tents, and bones to make tools.
- The South East had fertile land and a warm climate. So the tribes there settled down to farm. They built large towns, and developed a flourishing civilization.

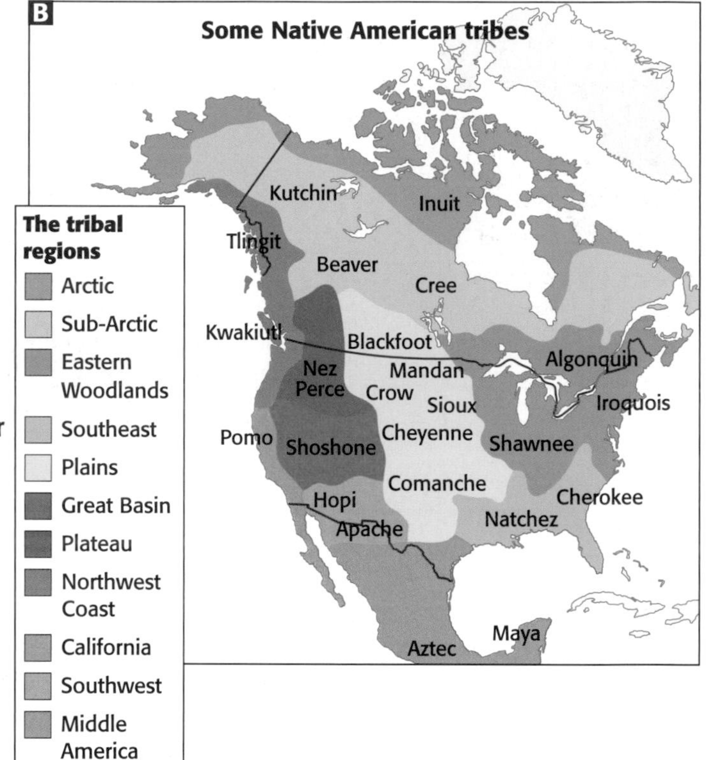

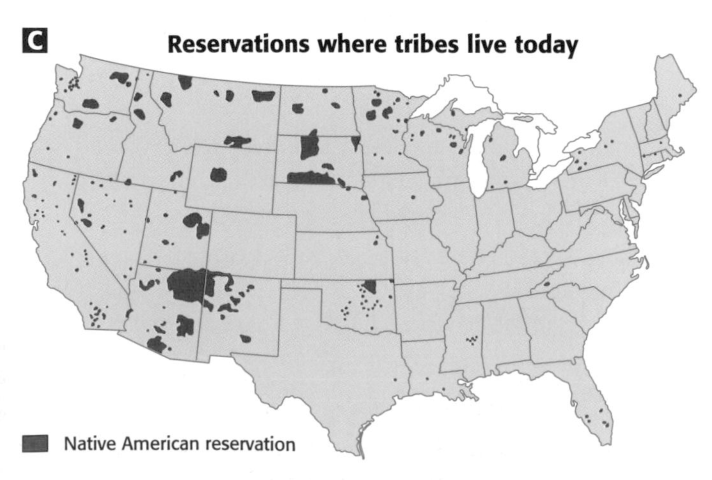

◀ *Chief Four Bears of the Mandan tribe, painted in 1832 by an American painter called George Catlin.*

When the Europeans arrived

We don't know how many Native Americans there were, when the Europeans arrived. But at least 2 million, and perhaps over 10 million.

At first, the Europeans and Native Americans were quite friendly.

- Tribes traded furs, leather, and food with the Europeans, in exchange for things like guns, knives, kettles, beads, and alcohol.
- The Europeans in turn learned about new crops, and how to farm them. For example corn, potatoes, pumkins, and beans.

Some Europeans were deeply impressed by the Native Americans and their way of life. Others treated them as little better than animals.

▲ *Native Americans on horseback. Many were very skilled riders.*

The American Indians today

A river of tears

When the white men came, it was the start of our river of tears.

First, the illness. They brought strange new diseases, that we could not treat with our native medicines. Smallpox was the worst. Many thousands of us died from it.

Then they took our land. We fought many battles with them over land. We fought bravely. But in the end, they won because they had more guns.

Some of them would have liked to kill us off, like vermin. Instead they set up reservations – land they permitted us to live on. Much of it was poor land. For example in the 1830s, they forced five tribes from their fertile farmland in the South East to a reservation on the Great Plains. Many thousands starved to death, on the journey and on their new poor land.

By 1900, we had given up the struggle with the white man.

▲ *A Native American family at home, on a reservation in South Dakota.*

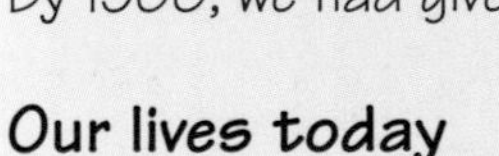

Our lives today

Today there are about 2.5 million of us. But we are no longer proud and free.

About half of us are scattered across America, in towns and cities. The rest live on reservations. There are around 300 of these. We run them ourselves. But the truth is, most of us live in great poverty. We have poor education, and little work. Many of us are alcoholics. Suicide is common.

Some settlements run casinos, where outsiders come to gamble. It brings in money. But it is also destroying the remains of our culture.

Did you know?

- *The Native Americans hunted eagles and wild turkeys, and used the feathers on arrows, and in their headdress.*

Your turn

1. Once upon a time, people could walk from Asia to North America. Explain why.
2. Look at map **B** on page 58. Which of these Native American tribes had you heard of before (if any)?
3. Look at Chief Four Bears. What do you think his clothing is made of? Map **B** and the text have clues!
4. Look at the Native American reservations, on map **C**. Where in the USA are most of them?
   a in the western half **b** in the eastern half
5. See if you can suggest reasons for the pattern you found in **4**. Maps **A** and **B** in Unit 3.1 may help.
6. So, what was the impact of the European settlers, on the Native Americans? Answer as fully as you can. It may help to compare maps **B** and **C**, on page 58.
7. *The Native Americans, and the tribes of the Amazon rainforest, have quite a lot in common.* From what you know already about rainforest tribes, and their history, do you think this is true? Explain.

3.5 The American Dream

What is the American Dream? And is it only a dream? You can find out here.

What is the American Dream?

The American Dream is the belief that, in the USA, everyone has the chance to succeed – to become and do what they want – no matter what their background is.

Over the centuries, this belief has drawn people to the USA from all over the world. And it is still doing so.

Only a dream?

The American Dream means different things to different people.

For some, it means a chance to live in freedom, and peace, and safety. For many, it means a comfortable home, and a good standard of living, and a good education for their children.

But whatever the dream, it has come true for millions of Americans and new immigrants. Here's a typical story.

▲ *Someone's American Dream?*

Their American Dream

My dad's from Mexico. He came here when he was just 16, with not a cent in his pocket, and only the clothes on his back. He took any work he could get. He swept streets, and cleaned offices, and worked in a bakery.

And then he started a little business, selling tacos and sandwiches in offices. That took off, and after a couple of years he opened a cafe. Then another, and another. Then he met mum, also from a Mexican family. She was working in an ice-cream parlor. So she joined him in the cafe business. She's always telling us how hard they worked, 7 days a week.

Now they have a lovely house, and two cars. They go on holidays to the Caribbean. They are so proud that I'm at law school, and my sister is an accountant. They saved really hard, to help us through college.

But yes, you could say their American Dream has come true.

Did you know?

- *In the USA, it's a basic principle that everyone has the right to life, liberty, and the pursuit of happiness.*

What if ...

- *... there were a British Dream? What would it be?*

What's the evidence?

Development indicators, and other data, help you to build up a picture of a country.

They give you an idea of what the people's standard of living is like.

Look at the data in this table.

How does the USA compare with the UK?

How does it compare with China?

Indicator	USA	UK	China
GDP ($US PPP trillion)*	14.6	2.3	7.8
GDP per capita ($US PPP)	48 000	37 400	6100
Life expectancy (years)	78	79	73
Under-5 mortality (per 1000 live births)	7	6	27
Doctors per 100 000 people	256	230	106
CO_2 emissions per person (tonnes)	20.6	9.8	3.8
Patents granted per million people	244	62	16
Murders per year, per million people	56	21	21

* A trillion is a billion billion.

How did it get to be like this?

Since the first Europeans arrived there, about 500 years ago, the USA has become one of the world's most wealthy and developed countries. In fact it has the world's highest GDP. How did it get to be like this?

First, the USA is rich in natural resources: fertile farmland, forests, rivers, metal ores, oil, gas, and the world's largest coal deposits. (But the oil is beginning to run out.)

Second, it started to develop industry long ago, soon after the UK. By 1900 it was the world's top industrial nation. Exports from its factories helped to make it wealthy.

Third, many of its immigrants over the centuries brought great skills. And most were prepared to work really hard, to make life better. Their attitude was 'We can do it'.

What about today?

Today, the USA's natural resources, and industry, are still important.

But much industry is in decline. Many companies have moved their factories to countries with lower wages. So, many American factory workers have lost their jobs. The USA now imports far more goods than it exports.

Today, most people work in the services sector (just as in other MEDCs). The USA is a world leader in areas like finance, and computer software. It is also a leader in scientific research: for example into human and plant genes, and new medical drugs, and all kinds of technology.

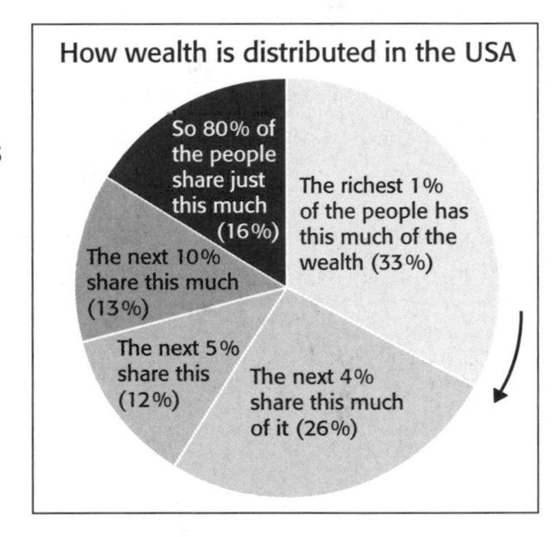

Not everyone is a winner

Not everyone in the USA is well off. There are millions of poor people. Around 13% of the population lives on less than $11 a day (about £7.70), for everything: rent, electricity, transport, clothing, food ...

Your turn

1 See if you can sum up the American Dream in:
a a ten-word sentence **b** a three-word slogan

2 These are terms from the table on page 60. See if you can explain them *without* looking in the glossary.
a GDP **b** GDP per capita **c** life expectancy
d under-5 mortality **e** CO_2 emissions

3 a One indicator in the table is linked to the amount of driving people do, and electricity they use. Which one? Explain why you chose it.
b What does this indicator tell you about life in the USA, compared with the UK and China?

4 One indicator suggests that the USA is a lively place for research, and new inventions. Which one?

5 a For which indicators in the table does the UK score *better than* the USA? List them.
b For each one, see if you can suggest reasons.

6 This table shows how people are employed, in the USA.
a Name three jobs, for each sector.
b Now write a paragraph about employment in the USA. See if you can make it interesting. Try terms like *most, fewest, over _ times more than.*

Employment in the USA

sector	% of workforce
primary	3%
secondary	19%
tertiary	78%

7 Now look at the pie chart above. Overall, what does it tell you about society in the USA?

8 So, do you think the American Dream is a reality?

3.6 I love my car!

This unit is about the car industry, which has been very important to the USA.

The dream on wheels

The stories of America and the car are closely linked. For many people, owning a car was part of the American Dream. And the bigger the better!

▲ *Two American icons: Elvis Presley, and his pink Cadillac.*

The USA's top manufacturing industries

petroleum (oil refining)
steel
motor vehicles
aerospace
telecommunications
chemicals
electronics
food processing

▲ *Out and about in a Ford Model T.*

It started with Mr Henry Ford ...

Henry Ford did not invent the motor car. But he was the first person to set up a factory making low-cost cars that ordinary people could afford. His factory was in Detroit, in the state of Michigan.

He started with the Ford Model T. It was launched on 1 October 1908. Over the next 20 years, 15 million Model Ts were sold, across America and around the world.

Why Detroit?

Detroit was a good place to set up a car plant, because it was in the area people called **the manufacturing belt**. This area had:

- lots of steel works, thanks to coal from the nearby coal fields, and iron ore from around Lake Superior. These were transported by water.
- plenty of workers, because most new immigrants arrived in New York, the USA's biggest city.
- lots of customers nearby. It was (and still is) one of the USA's most densely populated areas. Look at the map on page 57.

But Mr Ford was a fan of globalisation too. Within 20 years of opening his Detroit factory, he had set up car plants in the UK, Canada, France, India, Germany, and Australia. (Ford still has several plants in the UK.)

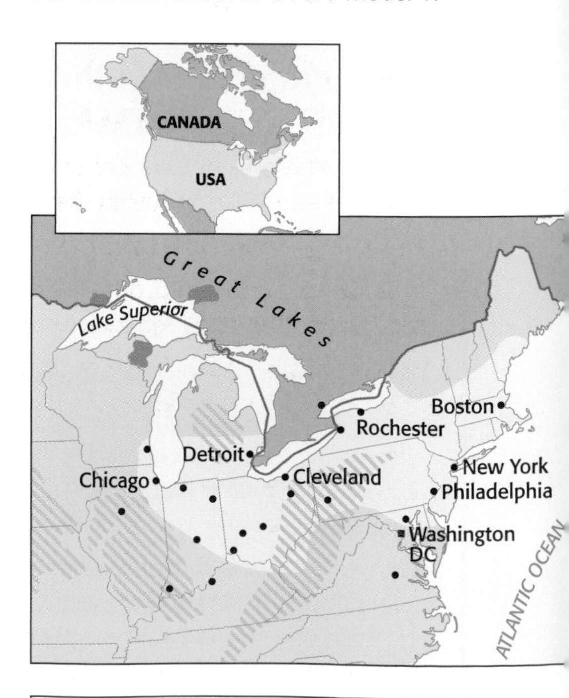

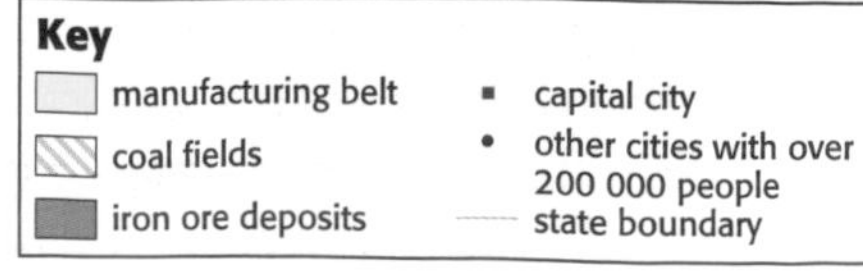

The impact on the American way of life

Other car companies learned from Ford. And soon, millions of people in the USA were proud car owners. They began driving everywhere, instead of taking trams. The impact was enormous.

- Now people could live further from their work. So new housing estates were built beyond the edges of cities. This encouraged **urban sprawl**.
- Large areas of cities were concreted over, to make car parks.
- Shopping malls with huge car parks were set up away from city centres, which caused many city centres to decline.

Today, in many suburbs, if you don't have a car, you can't easily get to a shop to buy basic things like milk.

▲ *Heading into Los Angeles.*

The bigger the better

For several decades, the American car industry flourished. Cars got bigger and flashier, with fewer and fewer miles to the gallon. But nobody cared. The USA had plenty of oil, so petrol (gas) was cheap.

All things must change

In the 1970s, the rosy picture began to change.

- First, the USA's oil deposits were being used up fast. So more and more oil had to be imported.
- In 1973 the world oil price shot upwards. So petrol prices rose too.
- Japan began to export smaller, less expensive cars to the USA. People liked them. So Japan opened car plants in the USA.
- Now there is pressure to come up with cars that don't burn petrol, to help fight global warming.

The car industry today

Today, the American car industry is in trouble.

- The Big Three American car companies – Ford, General Motors, and Chrysler – are still based in Detroit. But selling fewer and fewer cars.
- That's largely because of competition from foreign car companies.
- In 2008, the USA (and other rich countries) went into a period of **recession**. Car sales fell everywhere. So that did not help.

The government had to lend the Big Three money, to stop them going bust. But if they are not able to revive their business, they will have to close.

▲ *No need to get out!*

Your turn

1 Using the maps on page 62 and 139 to help you, say where the USA's main manufacturing belt is. Use geographical terms like *north* and *coast* in your answer. And name some cities and states!

2 a See if you can give two positive ways in which Mr Henry Ford helped to change the USA.

b Can you think of any negative impacts he has had?

3 *The car has helped to shape America's settlements, and change the landscape.*
Do you think this statement is true? Give evidence from the text. One of the photos may help too.

4 The Ford Motor Company is in difficulty. Imagine you are the head of Ford. What will you do, to help Ford survive? Come up with as many ideas as you can.

3.7 What about the farmers?

This unit is about farming in the USA. You might find some surprises!

▲ *Big field, big combine harvesters!*

▲ *Cowboys round up cattle on a Texan ranch.*

Not just big cars and movie stars

When you think of the USA, you might think of Disney, and American football, and jeans, and Coca-Cola, and big cars, and movie stars.

Now you can add farming! Just think about this:

- The USA is the world's top exporter of food (followed by Canada, then Brazil).
- Less than 2% of the American workforce is in farming. But overall, those farmers are the most productive in the world.
- You'll find almost any crop growing somewhere in the USA, thanks to the range of climates. Like cotton, oranges, tobacco, grapes, and rice. But the two main crops are wheat and corn.
- Rearing livestock, and especially cattle, is also big business.

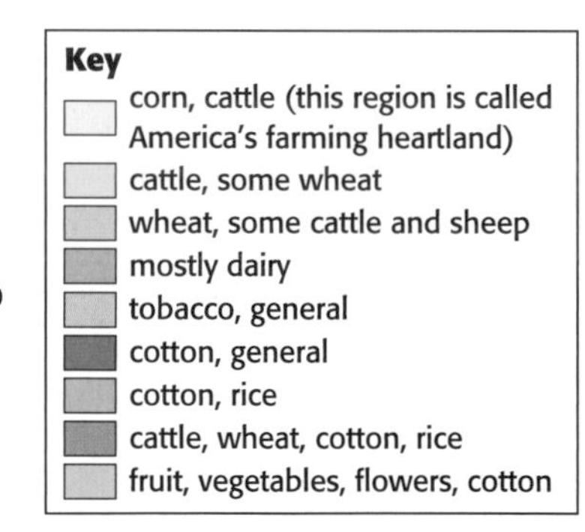

What's farmed where?

This map shows just the *main* farming activities in different regions of the USA.

It's a big country. And different crops ripen at different times of year. So some people earn a living by moving around with their combine harvesters, cutting crops for the farmers. Look at the first photo above.

For example a team might start in May, cutting wheat at **X** on this map. It may work its way up to **Y**, cutting any grain crops that are ready. And then turn around and work its way down again, until November.

It's hard work. You are on the road for months. You sleep in caravans. You might harvest through the night, if rain is on the way. But people really enjoy it.

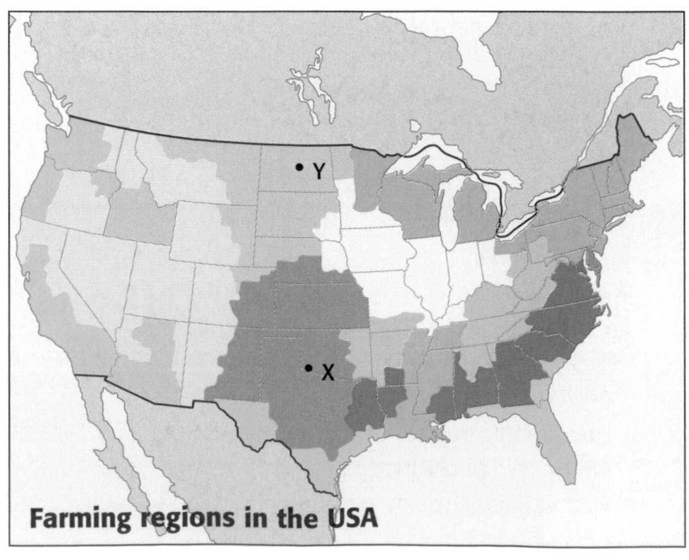

Farming regions in the USA

Why are they so productive?

Why are the USA's farmers so productive? Here are some reasons.

- **Good farmland** The USA has plenty of fertile flat land, good for growing crops and for using large machinery.
- **Large mechanised farms** Some farms are small, and produce a mix of things. But many are very large, and produce just one thing.

  For example, many wheat farms are over 4000 hectares. (Or more than 5600 football pitches!) They use big, modern, machinery. So a small number of workers can produce a lot.

  Some are owned by food companies, who then process the crops in their factories. Commercial farming like this is called **agribusiness**.
- **Chemicals** The farmers use lots of chemicals, such as fertilisers and pesticides, to increase crop yields and protect the crops.
- **Research** The USA leads the world in crop research. Scientists have been able to alter plant genes, to obtain plants that give bigger crops, or resist disease, or need less water.

  The farmers take advantage of this. Many now grow **genetically modified** (or GM) crops such as GM corn, rice, and tomatoes.
- **Subsidies** The US government also pays grants or **subsidies** to farmers for a number of crops, such as corn, wheat, rice, and cotton, and for dairy cattle. This encourages farmers.

▲ *A quick spray with pesticide.*

California: tops for farming

California is famous for Hollywood, and surfing, and Silicon Valley. It's also the top state for farming! (Find it on the map on page 139.)

California produces all kinds of fruit and vegetables, including grapes. It is big on cattle and milk. It also grows rice and cotton.

Water conflict

Crops need water. About 11% of America's cropland depends on irrigation. (Look at this photo.) The water comes from reservoirs that also serve towns and cities. In fact, about half the USA's fresh water goes to crops.

But in the dry western states, such as California, the population is rising fast. (12% of the USA's population now live in California.) So there is growing conflict over water. Should it be used for crops, or for people?

▲ *Not enough rain, so here comes the irrigation.*

Your turn

1 These big terms are used in the text. What do they mean?
  a subsidies **b** irrigation **c** mechanised
  d agribusiness **e** pesticides **f** genetically modified

2 Big mechanised farms, using plenty of fertiliser and pesticide, and lots of irrigation, can be very productive. The farmers can make lots of money.
  a But imagine you are the bird in the list on the right. How might you feel about these farms, and why? (The photos may give you ideas.)
  b Repeat **a** for each of the other entries in the list.

3 Now, from your own knowledge, see if you can identify any ways in which farming in the USA is:
  a similar to farming in the UK **b** different from it

Imagine you are …
- a bird that nests in hedgerows, and eats insects
- a mum who wants her children to have healthy food
- a refugee in Asia, living on rice donated by the USA
- a resident in a Californian city that's running short of water
- the US Minister of Finance, who keeps an eye on exports
- a farmer growing the same crops, in Ghana

3.8 California, the golden state

In this unit you'll learn about California, the USA's richest and most populous state.

The state that has it all

Each American state has attractions. But California has everything: coast, mountains, rivers, desert, big cities, fertile farmland, industry, a great racial mix, and tons of glamour!

It's the USA's richest and most populous state. Look at the factfile.

California factfile

capital:	Sacramento
largest city:	Los Angeles
wealth:	produces 13% of the USA's GDP
population:	37 million (or 12% of the USA's population)
pop mix:	42% white non-Hispanic 36% Hispanic 12% Asian 6% black 3.3% mixed race 0.7% Native American

A map of California

Here it is:

1 The climate

- Mild winters, and warm or hot summers, like the Mediterranean.
- But it's cold up the mountains. (You can ski there in winter.)
- The south of California gets much less rain than the north.

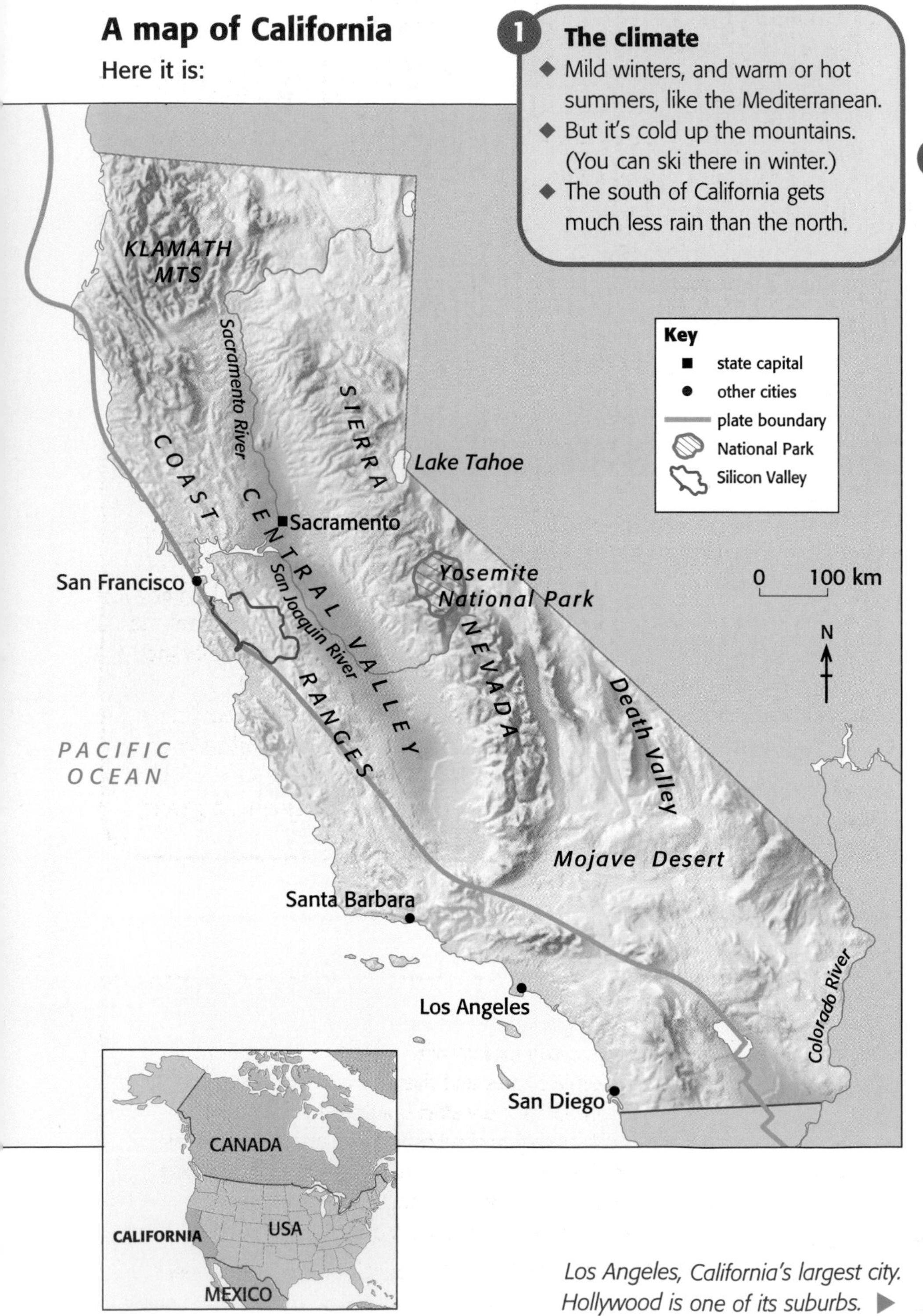

2 Physical features

- Look at the high Sierra Nevada mountain range.
- Look at the long, flat, sheltered Central Valley, drained by rivers.
- The map shows only the two main rivers: the Sacramento and the San Joaquin.
- Look at the Mojave Desert. Its lowest hottest part is Death Valley, which reaches 50 °C in summer.
- The coast has many beaches, from rocky and windy in the north to hot and sandy in the south.

▲ *A beach in southern California.*

Los Angeles, California's largest city. Hollywood is one of its suburbs. ▶

3 Farming

California is the top farming state.

- Most farming is in the Central Valley. Can you see why?
- They grow rice and cotton. And fruit, vegetables, and nuts – grapes, apricots, oranges, onions, tomatoes, almonds, pistachios …
- Dairy farming is important too.
- You'll see lots of food factories.

▲ *This farm is growing grapes.*

4 Industries

California has many industries.

- It has oil, on land and offshore. So oil refining is important.
- It is big on aerospace – all kinds of planes, and space vehicles.
- Find Silicon Valley on the map. This area is famous for computer and internet companies. Apple, Google, eBay, Yahoo!, and Intel are all based here.

▲ *Based in Silicon Valley.*

5 Movie time

- Los Angeles is the centre of the American film industry.
- Hollywood is one of its suburbs.
- Many movie stars live in Beverly Hills, an area within Los Angeles.

▲ *At the Oscars, in Hollywood.*

6 Tourism

California gets lots of tourists.

- Think of all those beaches, with surfing, sailing, diving.
- You can ski in the Sierra Nevada in winter and spring.
- Movie fans flock to Hollywood.
- The stunning scenery attracts millions of visitors – for example to Yosemite National Park.

▲ *In Yosemite National Park.*

7 Earthquake risk

- Look at the San Andreas fault on the map. It is the boundary between two plates: the Pacific and North American plates.
- The Pacific plate grinds along past the North American plate.
- So the rock is under great stress, and has developed many faults (cracks).
- So California suffers earthquakes. They predict a major one in the next 30 years.

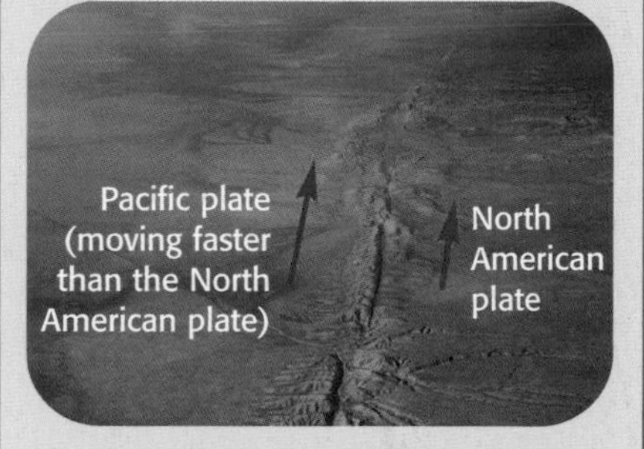

▲ *A view of the San Andreas fault.*

8 The water problem

California has a water problem.

- The north gets a lot of rain.
- The south is quite dry – but has far more people.
- Most of the crops need irrigation.
- So water gets pumped from the Sacramento River, to water the farmland of the south.
- There is conflict between the farmers and city dwellers over water use.
- Water shortages are a big worry.

Your turn

1 **a** Where in the USA is California?
  b Which other states border it? (Page 139.)
  c Which ocean lies off it?
  d Which country borders it?

2 California's Central Valley is one of America's top farming regions.
From the map, and the information on climate, see if you can suggest reasons for this.

3 See if you can explain *why*:
  a California's cities and rivers have Spanish names
  b California has very strict building regulations
  c you'll see big ads in Los Angeles about saving water
  d the Central Valley has a high % of migrant workers from Mexico, for some seasons of the year
  e very little rain falls in the Mojave Desert area.

4 Would you like to visit California? Give your reasons.

3.9 The USA: a superpower?

What's a superpower? Is the USA one? And what challenges does it face? This unit will help you answer those questions.

What's a superpower?

The term **superpower** is used to describe a country that has:

- a large population
- a large strong economy
- great military strength
- great political influence
- great cultural influence.

Is the USA one?

So is the USA a superpower? Read A – E, and see what you think.

Did you know?

- The US dollar is the main currency used in international trade.

Did you know?

- Of the world's 10 best-known brand names, 7 are American.
- No 1 is Coca-Cola.

A Population

- The USA has the third largest population in the world:

China	1300 million
India	1130 million
USA	304 million
UK	61 million

B Economy

- The USA is the world's largest economy. It produces more wealth each year than any other country. (It has the highest GDP.)
- Here's some GDP (PPP) data:

USA	$14.6 trillion*
China	$7.8 trillion
India	$3.3 trillion
UK	$2.3 trillion

* a trillion is a million million

C Military strength

- The world spends around $1200 billion a year on its armies, warships, fighter planes, and weapons.
- The USA spends by far the most – almost half the total!

▲ *An American stealth bomber.*

D Political influence

- The USA is big, strong, wealthy, and a military power. So it has plenty of influence on other countries.
- It usually wants to co-operate with them, to make the world a safer place.
- It gives aid to many poorer countries.
- But it has often used force, and the threat of force. So it has made some enemies.

▲ *The Great Seal of the USA.*

E Cultural influence

- American culture travels the world!
- It influences what people wear, and eat.
- We watch American films, and TV series, and listen to American music.
- Much of the influence is due to the spread of American companies like Coca-Cola and Nike.

▲ *American influence?*

But think about the challenges

Imagine you are the President of the USA. You are the world's most important leader. But it does not mean your life is easy. Look at the challenges you face.

1 Keeping America in work

- Many American companies are moving work to China and other low-wage countries.
- American stores are buying in cheap goods made in other countries, not America.
- All this means fewer jobs for Americans.

So how can you save jobs? Or create new ones? How can you keep the economy strong?

2 Immigration

The USA was built by immigrants.

- But now some people say you are allowing too many new immigrants in.
- They say it's putting a strain on housing, schools, and hospitals.
- Others say immigrants help the economy.

So should you carry on as you are? Or cut back on immigration? How will you get the right balance?

3 Oil dependence

Your country is addicted to oil. All those cars and trucks to feed! It's the world's top oil consumer. Around 20 million barrels a day.*

- The USA does have some oil of its own – but not nearly enough.
- So you spend a fortune every year on buying in oil.
- And there's that other problem linked to oil: global warming.

So how can you wean the USA off oil? How can you encourage new types of fuels, and cars?

* A barrel of oil is 159 litres.

4 Global warming

The USA is the top emitter of carbon dioxide per person. That gas is linked to global warming.

- They say global warming will bring you water shortages, and more powerful and frequent storms and hurricanes.
- The world wants the USA to cut its CO_2 emissions.
- But the USA depends on oil for transport, and coal for electricity.
- And cutting emissions will cost your industry a fortune.

How will you tackle this dilemma?

5 War, or peace?

The USA invaded Iraq in 2003. This has made it many enemies.

- Will you continue to use force against your enemies?
- Or will you do your best to make peace with them?
- Is your duty to make sure America is safe from enemies.

So what will your policy be?

Signing his first orders on fighting climate change, in 2009.

Your turn

1 a So do you think the USA qualifies as a superpower? Answer *Yes* or *No*.
  b Then give a set of bullet points to support your answer. Include data where you can.

2 For this, imagine you are an American teenager.
  a Look at A – E on page 68. From your point of view, which is the most important? Explain why.
  b List A – E in their order of importance, to you.

3 Do you think the President of the USA would give the same answer, for **2b**? Explain.

4 Do you think people in other countries would agree with the order you gave in **2b**? Give your reasons

5 Look at the challenges facing the USA. Which of them:
  a affect *only* Americans?
  b have an impact on people in other countries too?

6 Imagine you are the President of the USA.
  a Which challenge do you think is:
    i the most important to tackle? **ii** the most difficult?
  b Put the challenges in what you think is their order of importance, the most important first.

4 The global fashion

The big picture

This chapter is about **globalisation**. These are the big ideas behind the chapter:

- Globalisation means the way things are flowing more and more easily around the world. Things like goods, information, companies, jobs.
- Clothing is an example. Most of the clothes you buy are made in other countries – usually poorer countries.
- Their manufacture is usually arranged by big companies from richer countries, who want to make as much profit as possible.
- Globalisation brings many benefits. But it can also lead to people being exploited, or losing their jobs.

Country	Hourly wage in clothing factory in 2008 ($)
USA	11.16
UK	10.50
Mexico	2.54
Tunisia	1.68
Malaysia	1.18
China	0.86
India	0.51
Indonesia	0.44
Sri Lanka	0.43
Pakistan	0.37
Bangladesh	0.32

Your goals for this chapter

By the end of this chapter you should be able to answer these questions:

- What do these terms mean?
  globalisation *transnational corporation (or TNC)*
  revenue *profit* *GDP* *sweatshop*
- In what ways does globalisation affect my life?
- Why do companies like to go global?
- Some TNCs are more powerful than many countries. Why?
- Why do companies like Nike get things made in poorer countries?
- Many garment workers in poorer countries are exploited, to make our clothes. In what ways are they exploited?
- What are the aims of the World Trade Organisation (WTO)?
- What pros and cons can I give, for globalisation? (Give at least three of each.)

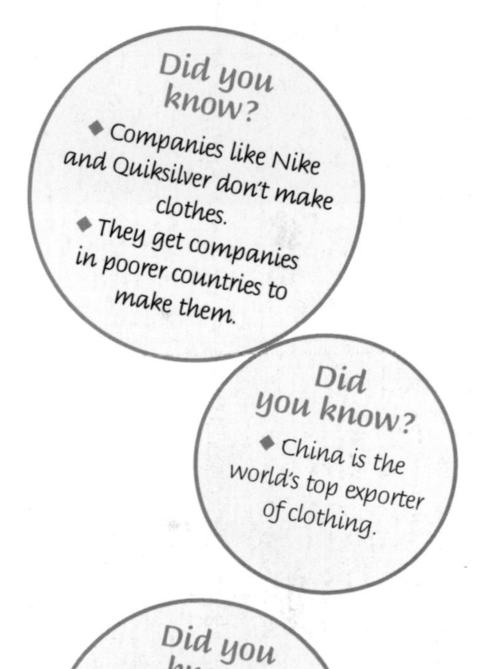

Did you know?

◆ The business of clothing humans is worth around 400 billion dollars a year.

And then ...

When you finish the chapter, come back to this page and see if you have met your goals!

Your chapter starter

Look at the clothes on page 70.

Do you think they might be expensive?

Who do you think made them?

Do you think they got paid much?

Why do clothes shops take a lot of trouble with their windows?

Our shrinking world

Here you'll find out what globalisation means, and think about how it affects you.

Compare these ...

In the 17th century, it took over 30 years for the French fashion for men's wigs to reach England.

Today, the latest fashion, whatever it is, can be copied around the world within hours.

In 1953, when Edmund Hillary conquered Mt Everest, it took 5 days for the news to reach London.

Today, you can have the final score for a sports event, on the other side of the world, within seconds.

100 years ago, if you wanted new clothes, you'd ask the local tailor or dressmaker to make them.

Today, the clothes you buy were probably made a few weeks earlier, in a country like China or Bangladesh.

It's globalisation!

The changes above are examples of **globalisation**.

Globalisation means the way goods, and companies, and information, and jobs, and fashions, are flowing more and more easily around the world. It's as if the world is shrinking! Some say we live in a global village now.

The term globalisation is used a great deal, these days. But note that it is *mostly* used about trade, and the activities of big international companies.

Explorers like Captain Cook helped to make globalisation possible. Here his ships Resolution and Adventure set off on his second great voyage (1771–75).

It started long ago

Globalisation started long ago. For example the traders who rode along the Silk Road 2000 years ago, carrying silk from China for the ladies of Rome, were part of the globalisation process.

Today, what's different is the speed of globalisation, and how many people it affects, and how much. You'll find out more later in this chapter.

It affects you!

Globalisation plays a big part in your life. It influences:

- what you wear, and the foods you eat
- what you do in your spare time
- your thoughts and opinions.

It will also influence what work you'll do as an adult, and how much you will earn, and perhaps even where you will live.

Any sign of globalisation?

Your turn

1 Say what *globalisation* means, in your own words.

2 **a** Choose at least three people from this list, and see how much you can say about them:
Barak Obama *Britney Spears* *Sachin Tendulkar*
Brad Pitt *Tiger Woods* *Angela Merkel*

b Do you think your answers for **a** owe anything to globalisation? Explain.

3 Now see if you give an example of:

a a sports event watched all over world

b a brand name that's known around the world

c a TV series watched, or copied, in many countries

4 80 years ago, almost all the food in our shops came from British farms. See how many foodstuffs you can list that we buy today, brought in from other countries. (Oranges are one example.)

5 Look at each item below. Do you think it has helped to speed up globalisation? If yes, explain why.

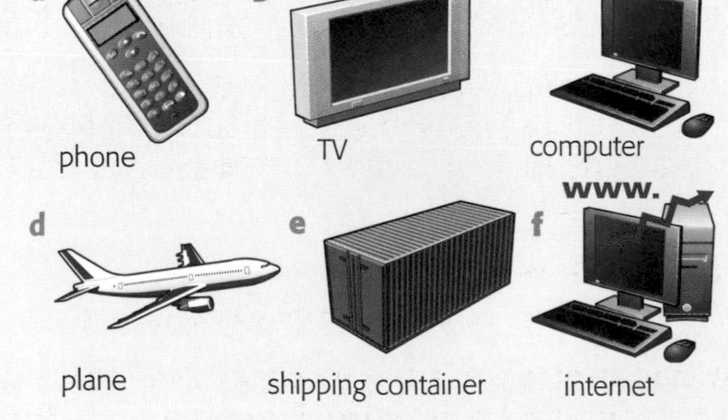

6 **a** Overall, do you think globalisation benefits *you*? Give reasons for your answer.

b Can you think of any harm it may do?

4.2 Walter's global jeans

Here you'll see how an ordinary pair of jeans can involve many countries.

> **Did you know?**
> - Denim was first made in the 18th century ...
> - ... for tough clothing for outdoor work.

A world of jeans

Jeans are a great example of globalisation – and not just because they are worn all over the world. Here's Walter, trying on his new pair.

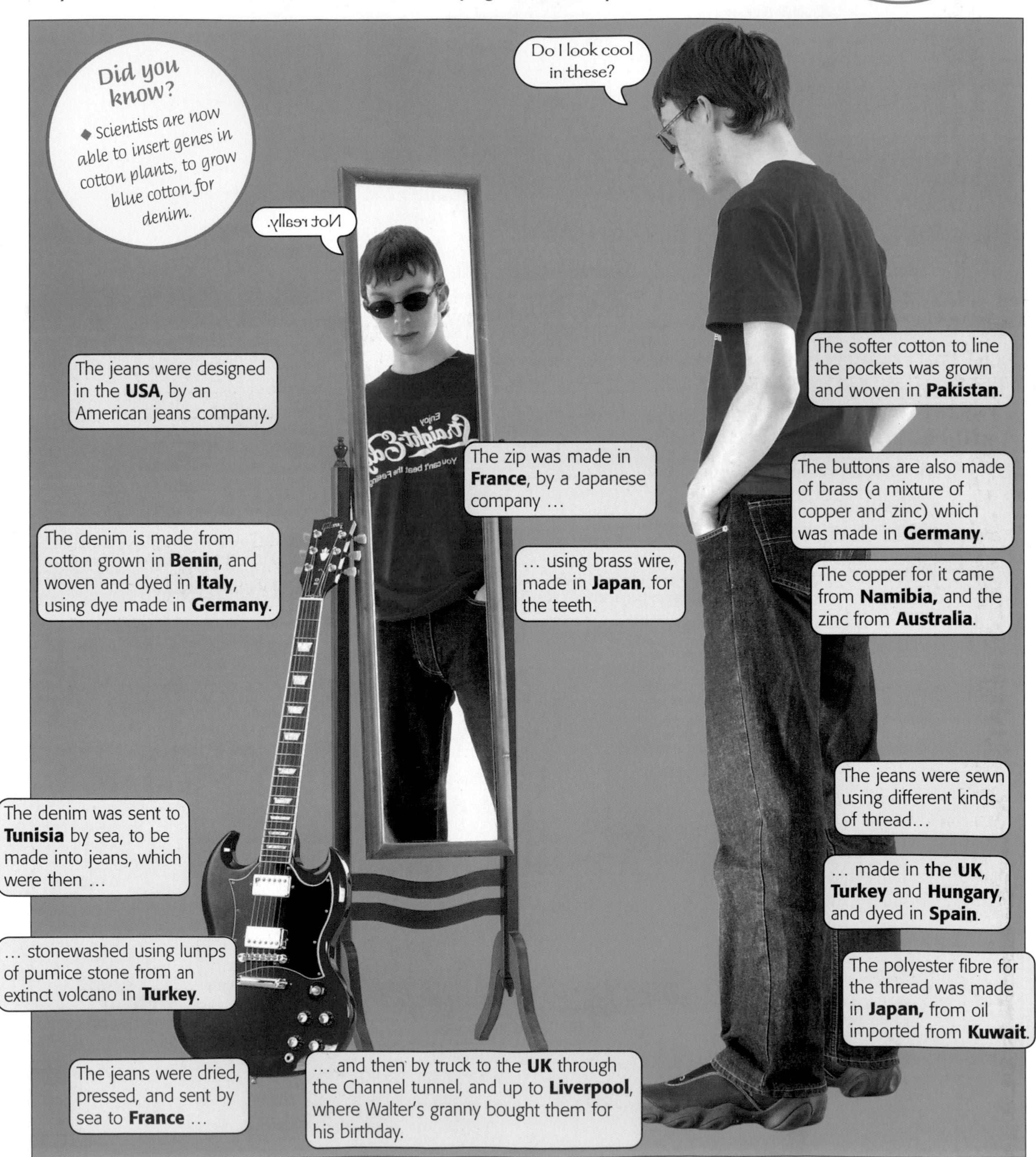

Getting Walter's jeans together

This map shows the 15 countries that were involved in producing Walter's jeans. Some of *your* clothes involve just as many countries. In fact you are all kitted out in geography!

Your turn

1 Which *continents* contributed to Walter's jeans?

2 **a** List the countries marked 1 – 15 on the map above. (See pages 140 – 141?) Answer like this: ① = ____

b Then, beside each country, write what it contributed to the jeans. (For example, *zip*.)

c Give your list a suitable title.

3 The map key is not complete. Complete it by matching the letters to the terms in italics below.
Give your answer like this: A = ____________
manufacture and processing of materials
source of a raw material
making and finishing the jeans
design and brand name
country where the jeans were sold
The glossary may help, if you're stuck.

4 Some countries on the map (for example the UK) have stripes of a second colour. Explain why.

5 The table on the right shows hourly wages in clothing factories. Use it, and the map above, to explain why:

a the American company didn't get the jeans made in the USA

b the denim was sent 1000 km from Italy to Tunisia, to be sewn into jeans.

Where your money goes when you buy a pair of jeans

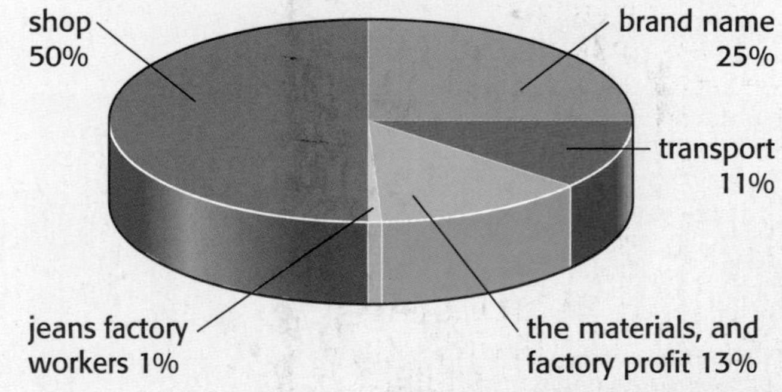

6 **a** Look at the pie chart above. Walter's granny paid £20 for his new jeans. Of this, how much went to:

i the shop where she bought them?

ii the American company whose label they carry?

iii the worker(s) who sewed them?

b Does what you found in **a** seem fair to you? Explain.

7 Do you think Walter's jeans had any impact on the environment? Give reasons for your answer.

Hourly wages for clothing factory workers

Country	Average pay per hour (£)
USA	6.64
UK	6.55
Italy	6.49
Tunisia	0.99

Nike: a global brand

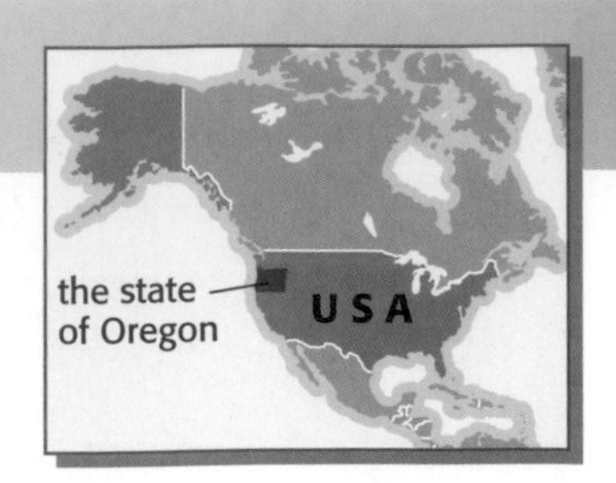

In this unit you'll see how Nike has spread around the world.

The Nike operation

Companies like Nike play a big part in the globalisation of trade – and fashion! Let's see how.

This is the Nike headquarters, in Oregon in the USA. Almost 6000 people work here. Many are busy thinking up new designs for clothing and footwear, to help Nike earn as much as possible from us.

Then Nike searches the world for places to get these things made, as cheaply as it can, in other people's factories. Nike goods are made in nearly 700 factories, in 52 countries. Mostly by young women like these.

While they are working hard, so is Nike – trying to get us to buy the things with the swoosh on. Nike spends over $2 billion a year on advertising, around the world. It pays top athletes millions to wear Nike products.

Nike goods are sent out to over 52 000 shops in over 180 countries – which means most countries of the world. It owns fewer than 600 shops itself.
Do any shops near you sell Nike?

The spread of Nike

This map shows how Nike has spread:

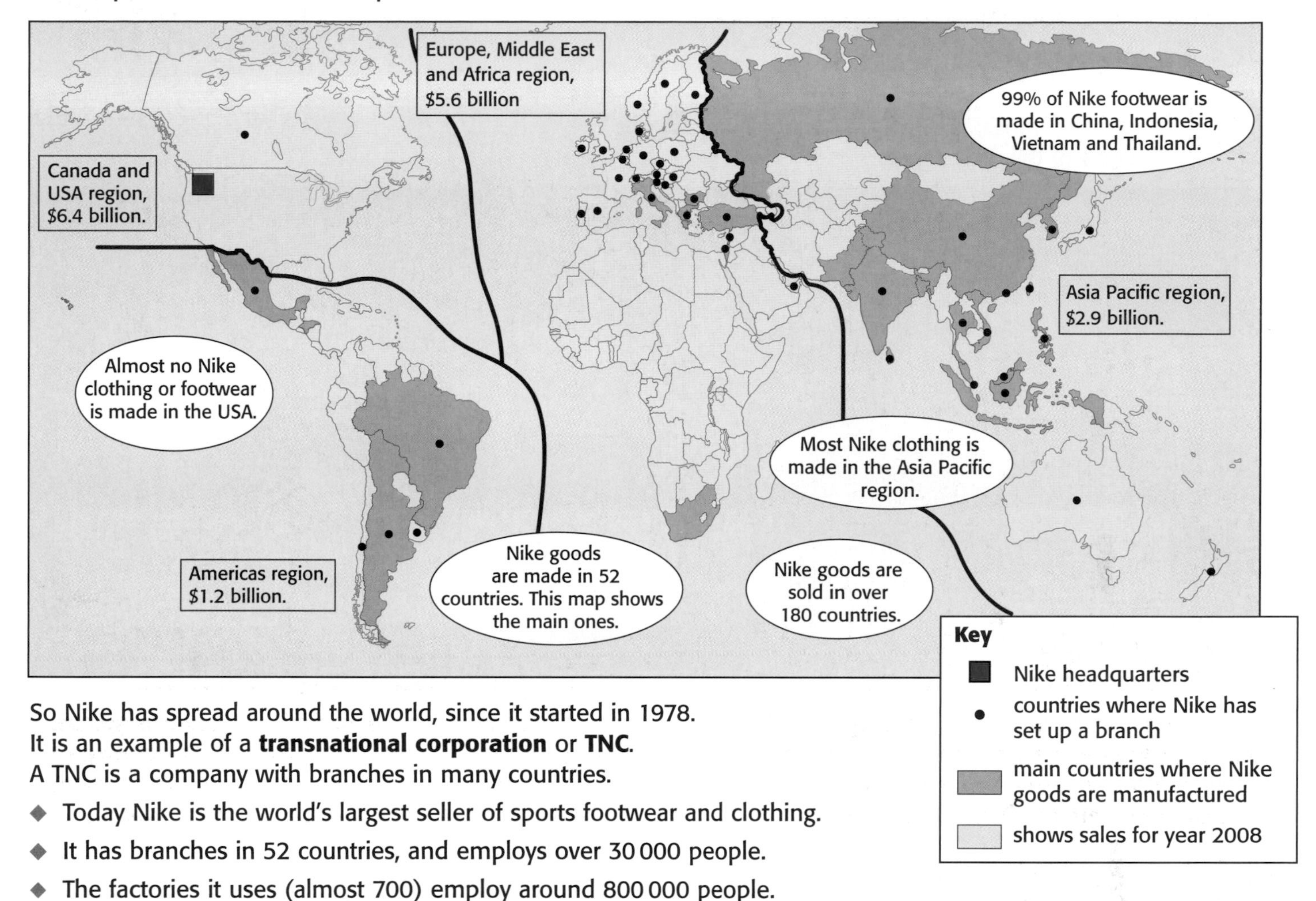

So Nike has spread around the world, since it started in 1978.
It is an example of a **transnational corporation** or **TNC**.
A TNC is a company with branches in many countries.

- Today Nike is the world's largest seller of sports footwear and clothing.
- It has branches in 52 countries, and employs over 30 000 people.
- The factories it uses (almost 700) employ around 800 000 people.
- It sells in over 180 countries (or nearly all the countries in the world).

Your turn

1 What kinds of goods does Nike sell? Write a list, and give prices if you can.

2 Nike spends over 2 billion dollars a year on advertising.
  a Why does it spend so much?
  b See how many sports stars or teams you can name, who are sponsored by Nike. (They wear the swoosh!)

3 Which of these is a better description of Nike? Explain.
  a a manufacturer b a design and marketing company

4 Using the map on pages 140–141 to help you, name:
  a six countries where Nike has a branch *and* gets goods made
  b two countries where Nike goods are made but Nike does not (yet) have a branch
  c eight other countries where Nike has a branch.

5 Compare the map above with the one on page 16.
  a Look at the GDP per capita for the countries where *most* Nike goods are made. What do you notice?
  b Suggest a reason why Nike chooses these countries.
  c On which *continent* does Nike *sell* most goods? Is it rich or poor?
  d Try to think of a reason why Nike does not (yet) get goods made in countries like Ghana.
  e Might Nike use Ghana one day? Give reasons.

6 Nike is a *transnational corporation* or TNC.
  a Explain the term in italics.
  b See if you can name any other TNCs.

7 *The spread of Nike is an example of globalisation.* Do you agree? Give reasons for your answer.

Why go global?

In this unit you will learn why companies like to spread around the world.

It's all about profit

Nike isn't alone. Thousands of companies have set up branches around the world to make things, or sell things, or both. So why do they go global? Because of this little equation:

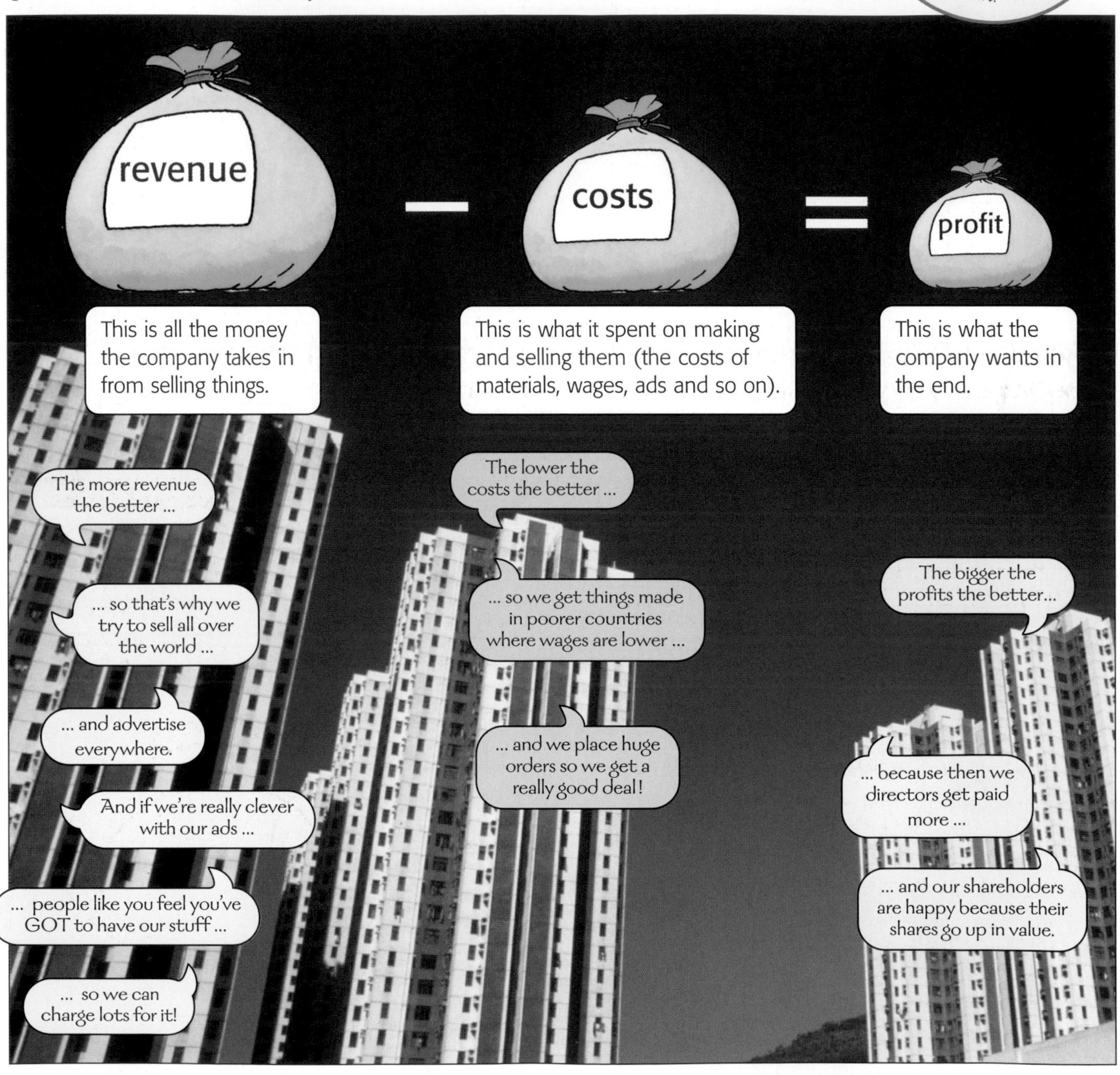

Did you know?
- In the UK, we spend around £49 billion a year on clothing and footwear ...
- ... or about £810 each.

So – does it work?

You bet! For Nike in 2008:

$18.7 billion	–	$10.2 billion	=	$8.5 billion (or $8500 million)
revenue		**costs**		**profit**

The US government then took a big chunk of this profit, as tax.

Did you know?
- Nike also owns Converse, Hurley, and Umbro.

A

Revenues for 10 TNCs in 2007			
TNC	**Its business**	**Based in**	**Revenue ($ billions)**
Exxon Mobil	oil/petrol	USA	390
Wal-Mart	supermarkets	USA	374
BP	oil/petrol	UK	292
Toyota	cars	Japan	265
Nestlé	foods	Switzerland	96
Microsoft	software	USA	58
Coca-Cola	you know what	USA	29
McDonald's	you know what	USA	23
GAP	clothing	USA	16
Nike	sports goods	USA	16

B

GDP (PPP) for 10 countries in 2007	
Country	**GDP (or total wealth produced) ($ billions)**
USA	13 780
India	2989
UK	2137
Belgium	376
Switzerland	300
Nigeria	293
Bangladesh	207
Tunisia	77
Ghana	31
Jamaica	21

Comparing TNCs

Look at table **A**. It shows 10 TNCs and their revenue, for one year. Note how small Nike is, compared with most of the others. The Wal-Mart supermarkets took in over 23 times more revenue than Nike, that year.

Comparing TNCs and countries

Some TNCs are doing so well, through globalisation, that they earn more than countries do.

Table **B** shows the GDP for 10 countries, for the same year as table **A**. The GDP is the total value of the goods and services a country produces. So Wal-Mart's supermarkets took in more revenue that year than Nigeria or Switzerland generated, for example.

In fact the combined revenue for the TNCs in table **A** that year was more than the combined GDP of 109 of the world's countries. Think about that!

▲ *Which TNC gets over 55 million customers a day, in 118 countries?*

Did you know?
◆ Wal-Mart owns the UK's Asda.

Your turn

1 Copy and complete, using terms from the brackets.
The more _____ a company sells and the _____ its _______ the higher its _______ will be.
(*money profits losses costs goods lower*)

2 Like every company, Nike aims to increase its profits.
a Make a *large* copy of the Venn diagram on the right. (Use a full page.)
b On your diagram, write in **A–H** below, *in full*, in the correct loops. (Small neat writing!) If you think one belongs in both loops, write it where they overlap.
A It gets 36% of its trainers made in China.
B It runs a website.
C It sponsors top school sports teams.
D It closed its trainer factories in the USA.
E It now owns no trainer or clothing factories.
F It employs sports scientists.
G It brings out new styles regularly.
H It opened a branch office in Australia.

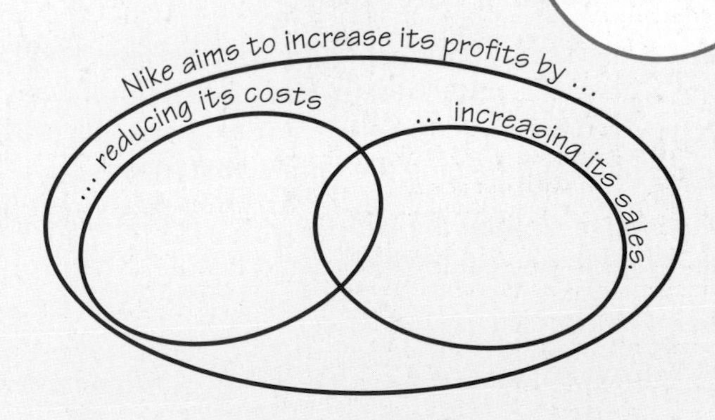

3 Now look at tables **A** and **B** at the top of the page.
a What is *GDP*? Give your answer as a sentence.
b Make *one* list showing the 20 companies and countries, in order of $ generated. (So USA first.)
c On your list, underline the countries in one colour and the companies in another. Add a key and a title.

4 Which do you think has more real power in the world, Wal-Mart or Ghana? Explain your answer.

A fashion victim?

Are people being exploited, so that we can look fashionable? Find out here!

It's just the fashion

It's not just Nike. Today, almost all the clothing and footwear on sale in our shops is made in poorer countries, where wages are lower. By whom? It could be someone like Shirin, in Bangladesh.

Shirin's day

It is only 9 am, but Shirin feels exhausted. She can hardly keep her eyes open. But if she stops sewing for even a minute the supervisor will yell at her. Or even slap her, like he did last time.

She's tired because she worked overtime last night. Until 2 am. She didn't want to, but if you refuse they sack you on the spot. Everyone is forced to do overtime now, for that big urgent order from the UK.

She dragged herself in again this morning, at 7 am. To sew more of those shirts. On and on, non-stop. By 8 her shoulders were aching. At 9, the heat in the factory is already stifling. And still an hour to go till the toilet break, when she can escape for 10 minutes.

Overtime again tonight. At 2 am she will leave the factory and hurry down the lane and into the slum, to the shack she rents with five other girls. No running water, no inside toilet. She will hurry, because she's so scared of being attacked in the dark.

More work tomorrow. And then a day off. How much will she have made this month? 1670 taka (*£14*) for four weeks, 48 hours a week. Then overtime pay on top of that. Maybe another 250 taka. But sometimes they cheat you on overtime – and you can't prove it.

By the time she pays the rent, and buys food, there won't be much left. But her mum is depending on her to send money home.

She thinks sadly about her home in the village, and her little brother and sisters. She misses them. Her mum did not want her to come to Dhaka to work. Her life here is not easy. But at least she's earning something, and helping the family. That means a lot.

She wonders about the people who buy the shirts. What are their lives like? And if they could see her and her life, what would they think?

Based on garment industry reports, 2008

Large or very large % of clothing made in low-wage countries*

Primark	Nike
Gap	G–Star
Topshop	Calvin Klein
M&S	Tommy Hilfiger
Asda	Quiksilver
Tesco	K–Swiss

* This is not a complete list!

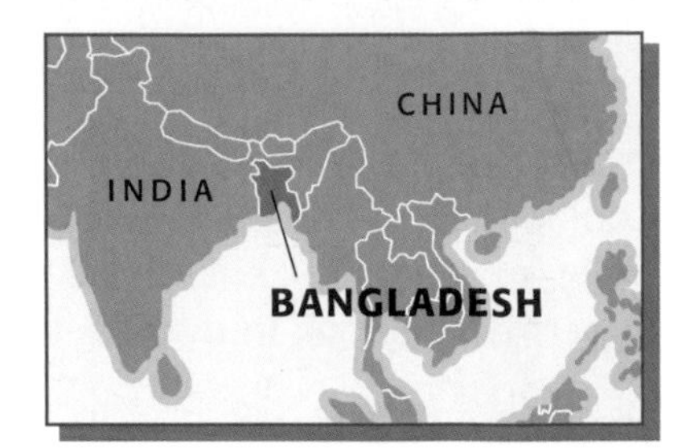

▲ *No rest for Shirin till she gets back …*

▼ *… to her rented shack in the slum.*

Just one of many …

You are connected to people like Shirin, through the clothes you wear. Many of the people who make our clothes, in other countries, are young women like her, from rural villages, with little education.

Some of the clothing factories are very modern. Some are dingy sheds. But even in the most modern factories, many garment workers do very long hours, for very little pay.

Clothing factories with long hours for low pay are called **sweatshops**.

How does it happen?

Why do people like Shirin end up working so hard, for so little? Let's see. We'll take a typical High Street clothing chain as example.

The catwalks are showing next season's styles. They're so exciting! Everyone will want them.

The clothing company rushes to have cheaper versions made. And that means heading abroad!

The factory owner wants the work, so he takes it on – and he aims to make a profit.

The clothes are needed quickly. So he forces his workers to work fast, for long hours – and very little pay.

The workers are not happy. But they need the jobs – even if they earn hardly enough to survive on.

The clothing gets finished on time. People like it. It sells really well. The company makes a huge profit.

Do you think it's fair?

Examples like the above are quite common. Not just for cheap clothing, but even for well-known labels that cost you a lot.

Clothing companies promise to make sure conditions are okay, in the factories they use. They say they check. But there are still many abuses. Is it fair that people like Shirin are exploited, to keep us looking good?

Did you know?

◆ *Primark is owned by Associated British Foods – a TNC that specialises in foodstuffs.*

Your turn

1 Shirin earns £14 for 4 weeks, doing 48 hours a week. How much is this: **a** per week? **b** per hour?

2 Make a list of the working conditions in Shirin's factory. You can put them in order, with what you think is the worst thing first. (Is low pay the worst thing?)

3 Now look at the chain above. What would happen if:
- **a** the factory owner refused to work for those terms?
- **b** the workers went on strike?
- **c** the government in the poor country passed a law that the factories must pay higher wages?
- **d** customers refused to buy clothes made in factories like Shirin's, because of the poor working conditions?
- **e** the clothing company decided to pay the factory more?

4 Of all the people in the chain, who do you think has:
- **a** the most power to change things?
- **b** the least power to change things?

5 **a** So, has globalisation changed Shirin's life? Explain.
- **b** If yes, do you think it has benefited her, or not?
- **c** Do you think she's being exploited? If so, by whom?

6 You are Shirin. What will you say to these people?

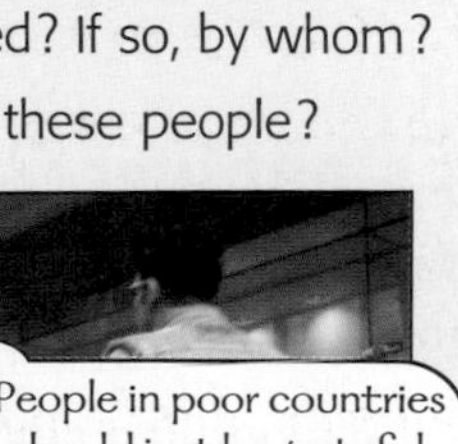

4.6 Global actions, local effects

other Dewhurst closures

Cardigan

In this unit you'll see how globalisation can lead to job losses.

The other side of the story

In Units 4.3 and 4.5, you saw how globalisation leads to jobs for people like Shirin. Now let's look at the other side of the story.

Why Kelly lost her job

This is Kelly at her machine. Or rather, it was Kelly at her machine. But the machine has gone. And Kelly's job went off to Morocco.

Kelly lives in Cardigan, in Wales. She used to work for Dewhirst, a big British clothing company, making jeans for Marks & Spencer. But then the factory closed.

Why did it close?

Most of Dewhirst's work was for Marks & Spencer. And that was the problem. M&S once got all its clothes made in factories in the UK. But by 1998, it was struggling to compete with other stores, who sold cheaper clothes made in poorer countries. Its sales slumped. So it decided to get its clothes made abroad too.

Today, Dewhirst still makes clothes for M&S – but in other countries! It closed its factories in the UK one by one. And set up new ones in Morocco, Indonesia, and Malaysia, where wages are much lower. So, like Kelly, thousands of workers in the UK lost their jobs.

Bitter sweet memories

'I loved working there', said Kelly. 'It was like a big family. But one day in 2002, the managers called us in and told us the factory was closing. We were shocked. Some women cried. Some had been there for over twenty years ! We had worked hard. But in the end that did not count. The company wanted cheaper workers.'

Regret

At the time, Dewhirst said 'We regret having to close this factory. But it all comes down to pressure from shoppers. They want cheap clothes !'

And still on the move

Dewhirst is still on the move. By the end of 2008, it had closed most of its Moroccan factories, and moved the work to China. Where next?

(Adapted from news reports)

▲ *Kelly at her machine.*

▲ *Workers leaving a Dewhirst factory for the last time.*

▲ *For over 100 years, M&S had all its clothing made in the UK.*

It's not just clothing ...

Clothing needs lots of people to sew it. And the skills are easy to learn. So poor countries want clothing factories, to help them climb out of poverty.

But it's not just clothing. Many other kinds of factories have closed in the UK, because production has moved to countries with lower wages. For example factories making cars, and mobile phones, and computer chips.

Your turn

1 Why did Kelly lose her job? The flowchart below will explain. But first you need to do some work on it!

a Make your own larger copy of the flowchart and drawings. (Just draw stick people.)
b Complete the sentences in the flowchart boxes.
c Walter's mother has a thought bubble. Draw bubbles for the others and fill them in, with thoughts about their part in the chain.

2 **a** Who may have lost out, when Dewhirst closed its UK factories? (Was it *only* the factory workers?)
b Who will have gained, when Dewhirst opened in Morocco? Think of as many groups as you can.
c Do you agree that Kelly was a victim of globalisation? Explain your answer.

3 Above are some ideas that a Dewhirst manager had, for saving Kelly's factory. Choose two. Say if you think they were good ideas, and give your reasons.

4 Textiles (cloth) and clothing were important industries in the UK for centuries. Now look at this graph:

a Is it true that the textile and clothing industries are in decline in the UK? (Glossary?) Give evidence.
b Suggest reasons for the trend the graph shows.

5

UK exports and imports of clothing (£ millions)

Year	1998	2000	2002	2004	2006
Exports	2480	2240	2180	2340	2590
Imports	5960	7200	8400	9140	10140

The table above shows exports and imports of clothing, for the UK (in millions of pounds).
a Show the export and import data on one graph. (Use any suitable type of graph.)
b Overall, what can you say about imports, compared with exports? Suggest a reason for this.
c Describe any trends you notice, and see if you can give reasons to explain them.

6 Now think up ideas to help the UK clothing industry. (You can't prevent imports.) For example, could it focus on 'specialist' clothing? (Bullet proof? Heat proof? Luxury?) Think about help for young designers too. Put your ideas in a memo to the Prime Minister.

So is globalisation a good thing?

In this unit you'll look at arguments in favour of globalisation.

It's going on everywhere

There is globalisation in all kinds of business, not just clothing and fashion. And it is mainly driven by TNCs. Is this a good thing?

1 The TNCs think so. (They would!)

2 Many governments in LEDCs think so.

3 Many governments in MEDCs think so.

4 Many workers all over the world think so.

5 Many economists think so.

6 The World Trade Organisation encourages trade.

So, many people are in favour of TNCs, and the globalisation of trade that TNCs promote. They think we all benefit. But as you'll see in the next unit, other people disagree!

More on the World Trade Organisation

Look back at point 6 on page 84. It mentions the **World Trade Organisation**, or **WTO**. This body was set up to help trade between countries flow more freely.

A key belief of the WTO is that global trade can help to reduce poverty, and promote peace and stability. Over 150 countries belong to it. Together, they try to work out a set of trade 'rules'.

The WTO aims to lower barriers to trade, so that countries can export goods and services more freely, within the agreed rules.

In past centuries, when countries fell out over trade they often went to war. Now, through the WTO, they can settle trade disputes peacefully.

Everyone here yet? Getting ready for a WTO meeting. ▶

Your turn

1 For each of these people, write down what you think is the *main* argument in favour of globalisation:

2 This is Vajra. She works in a call centre in Bangalore in India, for a British phone company.

Vajra is paid about 90 p an hour.
(Compare this to £6.73 an hour for the same job in the UK, and average pay of under 60 p an hour in India.)

a What's a *call centre*? (Glossary?)
b Is Vajra's work an example of globalisation? Explain.
c How many times more does a person in the UK earn per hour, for the same work as Vajra?
d What do you think the phone company would say, in favour of globalisation?
e What do you think Vajra would say, in favour of it?
f When it is 5 pm here it is 10.30 pm in Bangalore. What does that tell you about Vajra's working hours?

3 The UK is a member of the WTO.
a What is the WTO? Give its full name in your answer.
b Now see if you can write a short section for the WTO website, explaining how trade between countries can help to reduce world poverty. (120 words max.)

Against globalisation

In this unit you'll look at arguments against globalisation.

More harm than good ?

Many people are against at least some aspects of globalisation – and especially against the spread and power of TNCs.

1 Many politicians worry about the power of TNCs.

2 Workers in LEDCs have some criticisms.

3 Many workers in MEDCs have lost out.

4 Environmentalists have concerns.

5 Some economists are not too happy either.

6 Many people resent their culture being eroded.

Concerns about global trade

Most people agree that global trade can lift poor countries out of poverty. And the WTO is working to lower the barriers to trade. But:

- some feel that the WTO trade rules benefit the rich countries most, since these have the most power.
- some fear that TNCs help to shape the trade rules, by putting pressure on their governments.
- poor countries want to export as much as they can. But they'd like to restrict imports that could harm their own farmers, and industries. (There's more about this in Chapter 6.)

So talks on global trade are complex. But poor countries are starting to have more say. For example, through the **G20 group of developing nations**. (See page 141.)

Global protest

As trade goes global, so does protest.

The WTO holds a big meeting once every two years. Protesters from all over the world arrive. In the past, there has been violence. So meetings now tend to be in isolated places, with thousands of police on hand.

▲ *Get the message?*

Your turn

1 Page 86 shows things people say against globalisation. (These are mainly against the power of TNCs.)
Using page 86 to help you, give:
  a a *social* argument against globalisation
  b an *economic* argument against it
  c an *environmental* argument against it.
Give each as a short paragraph.

2 Globalisation is a complex issue.
  a From pages 84 and 86, pick out two arguments that are *exactly* opposite (one for and one against).
  b Now see if you can find *at least two more* pairs of opposing arguments. Write them down.

3 The photo above was taken at a big protest meeting against WTO trade rules and policies.
  a See if you can suggest three groups of people you might find at such a meeting. (TNC bosses?)
  b Study the message on the placard. What does it mean? See if you can rewrite it as a short speech.

4 On the right is Naresh, a security guard in India. He's guarding a building 4000 km away, in California! The CCTV pictures are sent by satellite. If he sees a problem, he raises the alarm.
You live in India. Write a letter to an Indian paper, in favour of, or against, the way the Californian company is employing Indian people.

5 Imagine you are the leader of a developing country. Foreign TNCs want to set up factories in your country, and provide services such as electricity, and build your new railways and roads. But you are a little worried about their power.
See if you can write some guidelines to help you decide about these TNCs. For example:
– should you try to protect local industries?
– what should you say about pay for local people?
– should your government get a share of the profits?
– how will you protect the environment?

6 But globalisation is not only about TNCs and trade. It's also about the spread of things like:
a knowledge **b** new medical treatments **c** news
Think about each in turn. Do you think everyone should be in favour of spreading this? Might anyone be against? Give reasons for your answers.

5 Coffee break !

The big picture

This chapter is about trade in coffee and other crops. These are the big ideas behind the chapter:

- Millions of small farmers in LEDCs depend on growing crops like coffee, for export.
- The crops are mostly bought by big companies, which process and package them for sale in our shops.
- The prices of crops on the world market change from year to year – and for many, prices have been falling for some time.
- Even in good years, the farmers don't earn much. Many live in poverty.
- Fairer trade rules would help poor countries escape from poverty.

Your goals for this chapter

By the end of this chapter you should be able to answer these questions:

- What kind of climate does coffee need, and which coffee-growing countries can I name? (At least five!)
- Who gets the smallest share of the money from a cup of coffee?
- The world price of coffee beans changes from year to year. Why? And how does this affect coffee farmers?
- How is Fairtrade coffee different from other coffees? And how does a Fairtrade company work with coffee farmers?
- What do these terms mean?
  subsidy *tariff* *World Trade Organisation (WTO)*
- Subsidies in rich countries can harm farmers in poor countries. What example can I give?
- Countries find it hard to agree on fair rules for world trade. What kinds of things do they disagree about?

And then ...

When you finish the chapter, come back to this page and see if you have met your goals!

Did you know?
- Coffee is the world's most valuable traded commodity, after oil!

Did you know?
- We think coffee was first discovered in Ethiopia ...
- ... at least 1200 years ago.

Did you know?
- The first known coffee house in England opened in Oxford in 1650.

What if ...
- ... everyone suddenly had a craving for coffee?

Your chapter starter

Look at the photo on page 88. The cherries contain coffee beans.

What do you think the children are doing?

Why do you think they're doing it?

Who do you think will drink the coffee?

Can you name any countries where coffee is grown?

5.1 Time for coffee

In this unit you'll learn how, and where, coffee is grown – and how the money you pay for a cup of coffee is shared.

Inside a cup of coffee

While you read this, millions of cups of coffee are being drunk all over the world. And hidden inside each one is months of toil.

1 Coffee grows on trees. The coffee berries are called cherries. They go from green to yellow to red as they ripen. In most places …

2 … the cherries are picked by hand. It's slow, because they ripen at different times! If it rains, the ripe ones get knocked to the ground.

3 Inside each cherry are two coffee beans. The beans are removed, washed well, then left to dry in the sun.

4 The clean dry coffee beans are poured into 60 kilogram sacks. The sacks will then be brought to a coffee centre.

5 There the beans are checked for size and quality. Some are rejected. The rest are packed again, ready for export by ship.

6 When they reach their destination, the beans must be roasted before use. Some are then crushed and processed to give 'instant' coffee.

Your turn

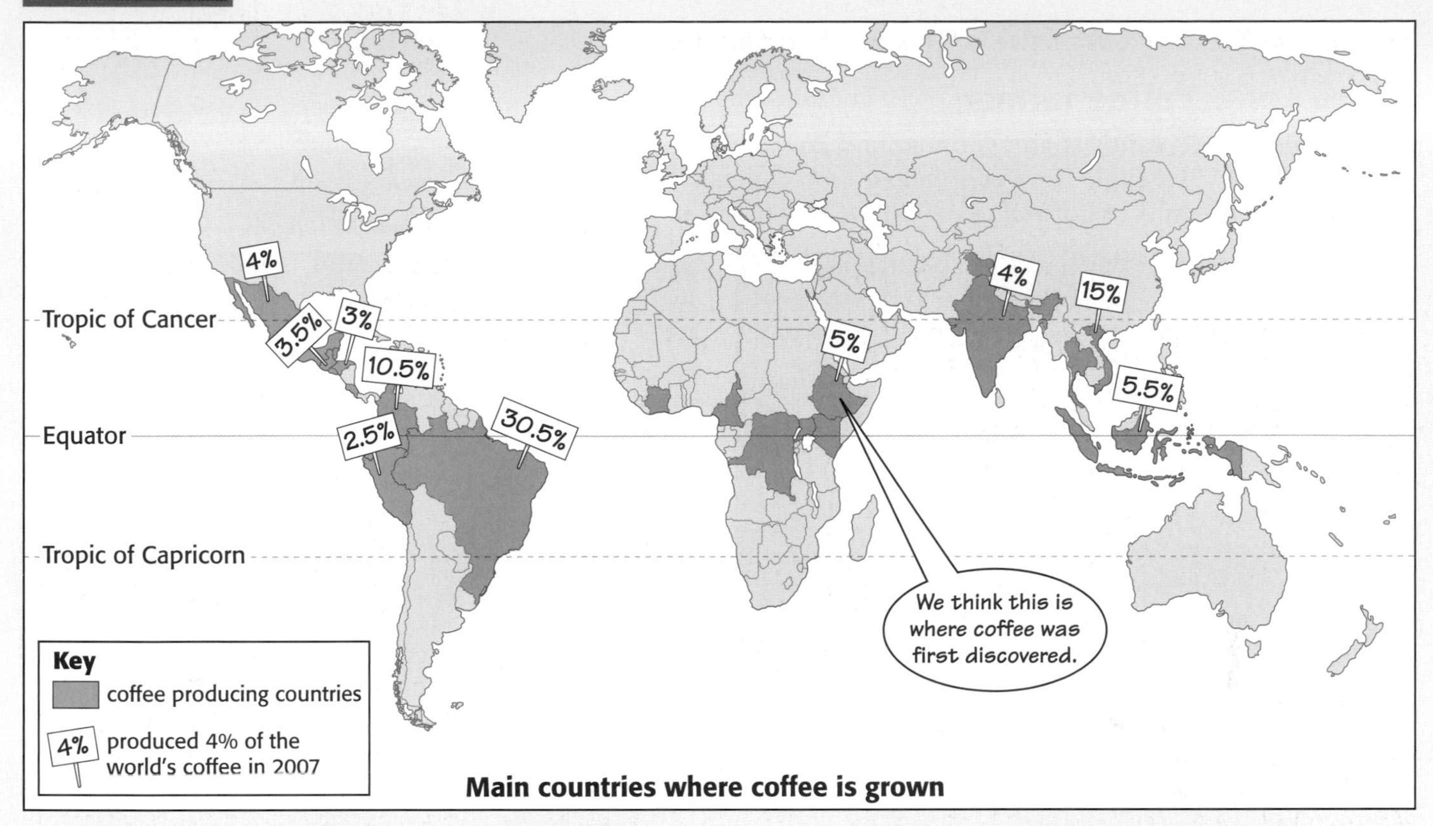

1 This map shows the main countries that grow coffee. It also gives the % share for the top 10 that year.

a List the top 10 coffee-growing countries and their % in order, the main one first. (Pages 140–141?)

b Now add one more row to your list to show the % grown by the rest of the coffee-growing countries.

c Next, draw a pictogram to show the data in your list. (Think of a good way to show 1% of coffee.)

d Which *continent* comes top for growing coffee?

2 Write down each statement. Then if you think it's false, cross out the wrong part and correct it neatly. (Page 90 and the map above will both help.)

A Coffee is grown in the tropics.
B Coffee trees need a cool climate.
C Russia depends heavily on its coffee exports.
D Dry weather is best for the coffee harvest.
E The Philippines is a top coffee producer.
F Ghana is the world's top coffee producer.
G They think coffee was first discovered in Tunisia.
H The coffee growing countries are all MEDCs.
I The coffee you drink will have travelled by ocean.
J The coffee industry is a global industry.

3 Look at the table on the right.

a Are there any coffee-growing countries in this list?

b Are there any LEDCs in this list? (Page 16?)

c What can you conclude from your answers for **a** and **b**? Give your reply as a full sentence.

4

This pie chart gives you an idea where the money goes, when you pay £1.75 for a coffee in a café.

a What % does the grower get?

b Who gets the largest share of the money?

c Most of the money ends up in ... ?
i the LEDC that grew the coffee **ii** an MEDC

d Where do you think most of the hard work is done?
i in the LEDC that grew the coffee **ii** in an MEDC

Top 10 coffee-drinking countries, 2003

Country	kg/person	Country	kg/person
1 Finland	11.4	**6** Switzerland	7.4
2 Iceland	9.1	**7** Netherlands	6.8
3 Norway	9.0	**8** Germany	6.6
4 Denmark	8.1	**9** Italy	5.7
5 Sweden	7.9	**10** Slovenia	5.6

Tricky coffee

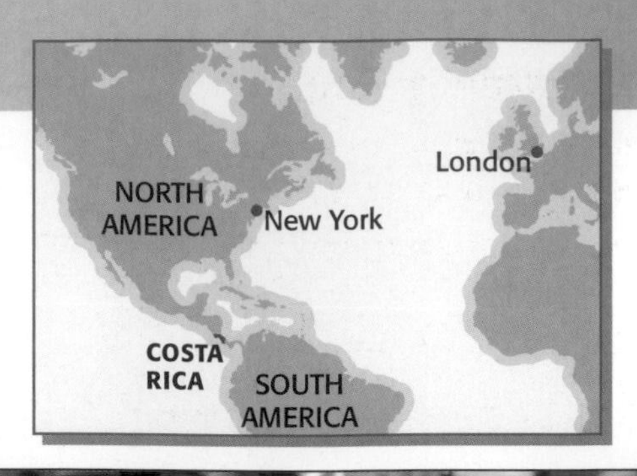

Here you'll find out why growing coffee for a living is a tricky business.

Pedro Loria, coffee farmer

Most of the world's coffee is grown on around 20 million small farms, in LEDCs. By people like Pedro below, who depend on it for a living. His farm is in Costa Rica, in Central America.

Growing coffee is hard work. Planting, watering in dry weather, spraying against disease, pruning, fertilising, picking. But even when Pedro works really hard, it does not mean he'll earn more. Because …

… the price is decided far away, in cities like New York and London, by people who'll never meet him. At a Commodity Exchange, dealers buy up the coffee crops, often before the harvest (mostly for TNCs like Nestlé).

Pedro does not know how much he will earn for his coffee, from one year to the next. That makes life difficult.

Another harvest over

It's the middle of February, and the coffee harvest has just finished. Pedro is worn out. For the last ten weeks he and his wife have started work at 5 am, and picked coffee cherries all day.

He has heard he'll get a better price for his beans this year. It is still not much. But he'll be able to afford the clothes and books and things his children need for school. Perhaps he can even save some money.

So he feels lucky. He thinks back to 2001, that awful year, when he earned so little. Only the vegetables they grew, and the chickens, kept his family from starving. Not like his neighbours, who had to go begging down the road.

It's a harsh world. His cousin told him how much people in London pay for a coffee in a coffee bar. A fortune! And the price goes up every year. So why does he get so little for his beans? That makes him feel bitter.

What about next year? He has no idea what he'll earn. But it could be much less than this year. And one year, coffee may break him.

Did you know?

- The coffee industry is worth over $80 billion a year.

Did you know?

- Four TNCs between them control 40% of the world's coffee: Procter & Gamble, Kraft Foods, Sara Lee and Nestlé.

The coffee mystery

More and more of us are drinking coffee. But even in a good year, Pedro does not earn very much from growing coffee. He remains poor. Why?

It's mainly because there's too much coffee grown! There are tonnes of beans stored around the world.

Why is so much grown? Partly because the World Bank and others have encouraged LEDCs to grow it.

So countries have competed to grow more and more, instead of agreeing a plan between them.

Then, with so much coffee on the market, the buyers can push the price right down. They pay less …

… so the coffee farmers earn less. Less money for food, and clothes, and education for their children.

Meanwhile more of us like to drink coffee in smart coffee bars. So they can charge us more for it.

Your turn

1 See if you can copy and complete this:

The more coffee on sale on the world market, the _____ the price will be. The _____ coffee on sale, the _____ the price will be.

2 How do you think this will affect the world price of coffee beans?
- **a** The coffee-growing countries produce bumper crops.
- **b** A disease destroys all the stores of coffee beans.
- **c** Coffee-drinking grows really popular in China.

3 Now look at the graph on the right.
- **a** What does it show?
- **b** What can you say about the shape of the graph line?
- **c** In which year was the price:
  i highest? How high? **ii** lowest? How low?
- **d** Imagine you are a coffee farmer, like Pedro. What problems might a graph line like this cause you? See how many you can think of.

4 Look again at the graph. The price of coffee rose between 2000 and 2007. Can Pedro be sure it will just keep on rising? Give reasons for your answer.

5 Pedro receives only about 3% of what you'd pay for a coffee made with his beans.
- **a** Do you think this is fair?
- **b** What do you think could be done about it?

The price of coffee beans on the world market

Price (US dollars per kilo): 0, 1, 2, 3, 4, 5, 6

Year: 1975, 1980, 1985, 1990, 1995, 2000, 2005, 2010

5.3 A fair price for coffee farmers

In this unit, you can find out about one way to help the coffee farmers earn a fair price for their coffee.

▲ *Many coffee farmers around the world can barely make a living.*

How the world market works

On the world market, the price of coffee depends on just two things:

- the **supply** of coffee – how much is for sale
- the **demand** for it – how much people are prepared to buy.

And while trading goes on, no-one thinks at all about the farmers, or how hard they worked, or much they need to earn to survive.

Another way to look at it

There's another way to buy and sell coffee, that keeps the coffee farmers in mind: *Decide on a fair price to pay the farmers.*

It is called **fair trade**. The flowchart below shows how it works.

It involves the **Fairtrade Foundation**, which was set up in the UK by Oxfam and other charities.

Fairtrade coffee

This is how fair trade works for coffee:

A coffee company and a group of coffee farmers decide to work together.

↓

They agree a fair price for the coffee, that gives the farmers a decent living – and add some extra money.

↓

The company pays some of the money in advance, before the harvest, so that the farmers won't run short.

↓

The Fairtrade Foundation has to check the deal, before the company can use the Fairtrade logo on its coffee.

↓

In the shops, people who want to help the coffee farmers will buy the coffee – and they don't mind that it costs a bit more.

▲ *If it has the logo, it's Fairtrade.*

In return the coffee farmers promise to treat *their* workers fairly, and to look after the environment. And they must use the extra money for projects that will help the whole community. For example to build a new primary school, or a well for fresh water.

How is it going ?

Sales of Fairtrade coffee are rising fast in the UK. And not just coffee. You will see the Fairtrade logo on tea, sugar, fruit, juices, chocolate, honey, roses, wine, cotton wool, cotton T-shirts, and even footballs ! More goods are on the way.

And it's not just the UK. Fairtrade goods are now sold in 19 countries.

▲ *Fairtrade is not just for coffee ...*

How Fairtrade helps the coffee farmers

The Fairtrade movement helps many thousands of coffee farmers.

Celia is one. She lives in Cameroon, in Africa. She and a group of other farmers sell their coffee beans to a British Fairtrade coffee company.

'Before, we had nothing,' said Celia. 'But now we have desks and chairs in the primary school, so the children don't have to sit on the floor. We pay for some children to go to secondary school. We repaired some bridges. And best of all, we built a clinic. So now a nurse comes to visit.'

'We are still poor. But life has improved so much already. You see, fair trade is all the help we need. We don't need charity!'

▲ *Hard work. But for fair pay, from a Fairtrade coffee company.*

But what about all the other coffee farmers ?

There are around 20 million coffee farmers. Fairtrade can help only a tiny fraction of them. The rest are at the mercy of the world trading system. Oxfam and other groups made these suggestions to help them:

You'll explore them in '*Your turn*.'

Your turn

1

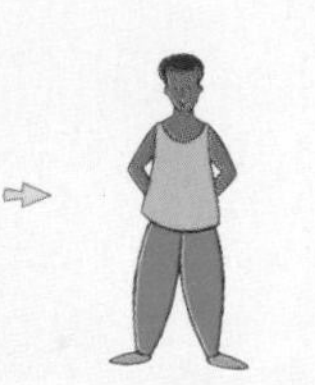

shopper → supermarket manager → from Fairtrade company → coffee farmer

The people in this chain are linked by fair trade.

a Make a large copy of the chain. (Draw stick people.) Give each person a *large* speech bubble.

b Write each of these in the correct bubble, and then complete the sentences:

By buying coffee beans directly, we ...
Now I can get on with growing coffee, without ...
I like to buy coffee that ...
We make a fair profit on the coffee – so ...

2 a Is there a loser, in your chain for **1**? If so, who?
b Is there a winner? If so, who ?

3 Look at how Fairtrade companies work with coffee farmers. Do you think it's *sustainable*? Explain.

4 Now look at suggestions Ⓐ – Ⓓ above.

a Ⓐ might seem a bit shocking. See if you can explain how it would help coffee growers.
b Which other suggestion could reduce *supply*?
c Which do you think would be fastest to carry out?
d Which might take longest? (At least a few years.)
e Which might be the most difficult to carry out? Why?
f Explain how Ⓓ would help coffee-growing countries. (Hint: pie chart on page 91.)

5 Now draw a consequence map like the one begun here, for Ⓓ. Add as many boxes as you can.

Coffee growers make and export instant coffee ...
↓ *which means*
... they earn more money ...
↓ *so*

6 You could give money to a charity to help coffee farmers – or buy Fairtrade coffee. Which do *you* think is better? Give reasons.

5.4 It's not just coffee

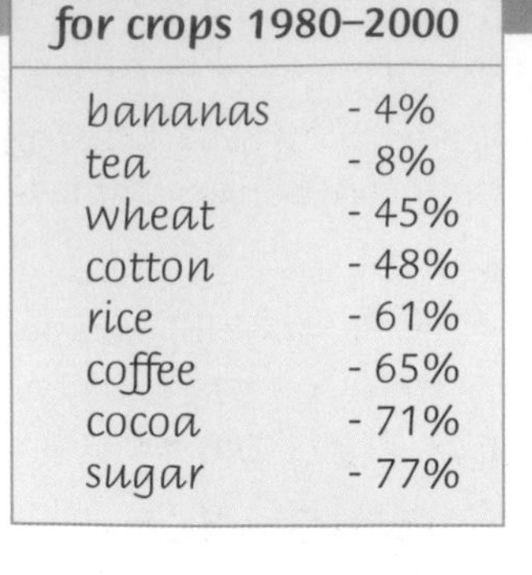

% fall in real prices for crops 1980–2000	
bananas	- 4%
tea	- 8%
wheat	- 45%
cotton	- 48%
rice	- 61%
coffee	- 65%
cocoa	- 71%
sugar	- 77%

Here we take a wider look at the trade in crops, and the effects of subsidies and tariffs. (You met those already on page 21.) We use cotton as an example.

The long slide down?

In Unit 5.2 you saw how the world coffee price changes from year to year. And for many years, the overall trend was down. But it's not just coffee. The prices of many crops have been falling. Look at the table on the right.

The usual reasons are:

More farmers in more countries start growing a crop. → Many also use better seeds, and fertilisers, and pesticides. → So there are fine big harvests of the crop, around the world. → So its world price falls. (Price falls as supply rises.)

Losers and winners

When the price of a crop falls, all the farmers are affected. But those in the poorest countries are the biggest losers. They sink deeper into poverty.

The winners are often the big TNCs, who buy tonnes of the crop cheaply. And if they sell us the goods more cheaply, we win too. (They might not!)

Did you know?

◆ Cotton is the world's top non-food crop.

The subsidy problem

For some crops, the poorest farmers also face other obstacles, over which they have no control. You are probably wearing some cotton right now. So let's take cotton as example.

Millions of African farmers grow cotton, on small mixed farms. These are cotton farmers in Burkina Faso. It is one of the very poorest countries in the world.

About 25 000 farmers in the USA also grow cotton. Some of their cotton farms are enormous. (Some were once plantations worked by African slaves.)

The American cotton farmers have one big advantage. The US government pays them a **subsidy** for growing cotton. The bigger the farm, the bigger the subsidy.

The subsidy encourages the American farmers to grow cotton. So they grow a lot. In fact the USA is the world's top cotton exporter.

But it hurts the farmers of Burkina Faso and the other African countries. Because the more cotton American farmers grow, the lower the world price.

Most of the African cotton farmers have very little to live on. A fall in cotton prices may mean hungry families. No money for medicine, or school fees, or clothes. No fertiliser for next year's crop.

Compare …	Burkina Faso	UK
population	15 million	61 million
GDP per capita (PPP) in $US	$1200	$35 000
life expectancy	51 years	79 years
adult literacy rate	24%	100%

It's not just cotton

The USA, and other richer countries, give subsidies for other crops too. For example the European Union gives £40 billion a year to farmers in the UK and other EU countries, to grow crops like wheat and sugar beet.

The poorer countries say this is not fair. Without subsidies less would be grown, and their own farmers would get better prices.

▲ *Cotton farmers in Peru protesting against cotton imports from the USA.*

Then there are tariffs

A **tariff** is a tax you may have to pay your own government, when you import something from another country. If it's high enough, it puts you off importing! All countries like to put tariffs on crops that their own farmers already grow. It's a way to protect the farmers from competition.

Who decides what's fair?

As you saw on page 85, the World Trade Organization (or WTO) was set up to help world trade flow more freely. Over 150 countries belong to it. They meet to negotiate on 'rules' for world trade.

But they find it hard to agree about crops. They argue along these lines:

We depend mainly on farming, and we're poor.

Fairer trade rules would help us climb out of poverty.

We, the developing countries, want you to:

- drop subsidies
- cut tariffs on crops we sell you
- agree to us putting tariffs on imported crops, when we need to protect our farmers.

We have farmers to look after too.

And this is trade, not charity.

We, the richer countries, want you to:

- open up, and let us sell our crops to you freely, without tariffs.

Else we won't drop subsidies, or cut tariffs.

Did you know?

- You may even have eaten some cotton!
- Seeds grow with the fibre, and some are crushed to give oil for cooking oils and salad dressing.

WTO trade talks about crops have been going on since 2001!

Your turn

1. Burkina Faso is in Africa. But where? Find it on the map on page 141. Then say where in Africa it is, and name the countries that border it.
2. *Burkina Faso is a highly developed country.* Do you agree? Give evidence to back up your answer, using the table at the bottom of page 96.
3. Cotton forms about 80% of Burkina Faso's exports. Do you think that's a good thing, or bad? Give reasons.
4. American farmers produce over twice as much cotton per hectare as Burkina Faso's farmers do. See if you can suggest reasons. The photos on page 96 may help.
5. The USA exports about 20 times more cotton than Burkina Faso does.
   - **a** What is a *subsidy*?
   - **b** What do you think will happen to the USA's cotton exports, if the subsidies are dropped? Explain why.
   - **c** What may happen to Burkina Faso's earnings, if the USA drops its cotton subsidies? Why?
6. **a** What are *tariffs*? Give one reason for having them.
   - **b** You are Minister of Trade, in Burkina Faso. Will you want to put a tariff on imports of cotton fibre? Give your reasons, in a speech for a WTO meeting.

6 Tourism – good or bad?

The big picture

This chapter is about tourism. These are the big ideas behind the chapter:

- We humans have always explored the world – and still do, through tourism.
- Tourism is a big important industry, employing a great many people.
- Many poor countries rely heavily on tourism, as a way to earn money.
- But tourism can exploit people, and damage environments.

Your goals for this chapter

By the end of this chapter you should be able to answer these questions:

- What do these terms mean?

  tourist *tourism* *sustainable tourism*
  domestic tourist *inbound tourist* *international tourism*

- International tourism has been growing fast over the last 50 years. Give reasons for this. (At least four.)
- Which is the most visited region for domestic tourism, in the UK?
- Which are the two most visited countries, for UK residents taking holidays abroad?
- What else can I say about holiday patterns in the UK? (Give at least two more facts.)
- What do these terms mean?

  package holiday *mass tourism* *ecotourism*

- Where is Benidorm, and how has mass tourism changed it?
- Where is Gambia, and what does it offer tourists?
- What are the benefits, and negative points, of tourism for Gambia?
- The Ese'eja project in Peru is an example of ecotourism. Why?
- Where in the UK are the Broads? What are they like? What kinds of activities do they offer, for tourists?
- Overall, is tourism good, or bad? What do you think?

And then …

When you finish the chapter, come back to this page and see if you have met your goals!

Did you know?

- *Tourism is the fastest-growing industry on the planet.*

Did you know?

- *In 2004, 760 million people visited other countries, as tourists.*
- *In 2024, the number could be 1.5 billion.*

Did you know?

- *Around 195 million jobs around the world are connected to tourism.*
- *That's about 8% of all jobs.*

What if ...

- *... flying was banned, to help fight climate change?*

Your chapter starter

The man and his camel are waiting, in the photo on page 98.

Who do you think they're waiting for?

What will happen if nobody turns up?

What country do you think this is?

Would you like a job like that?

6.1 Introducing tourism

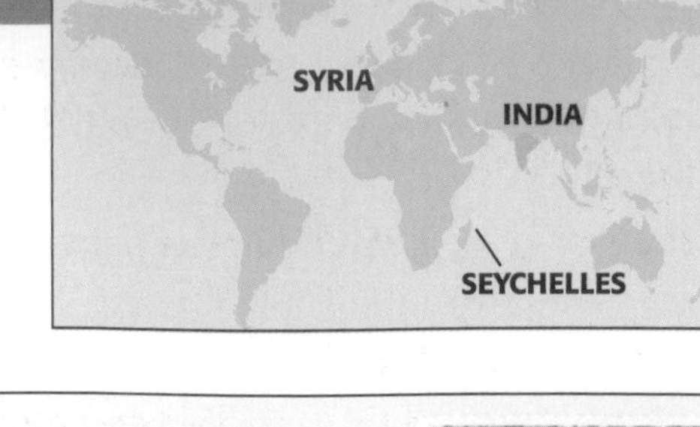

Learn here what tourists and tourism are, and the essentials for tourism.

Wish you were here?

At this very moment, all around the world, the post is bringing thousands of postcards from tourists …

A

This is my beach! My little chalet is on the edge of the sand, under a palm tree. I open the shutters in the morning and see this view.

Then coconut pancakes and orange juice for breakfast, and a lazy day swimming and reading. Bliss.

X Susie

Joe Finn
9 Bar Lane
Linden CB25 3RU
United Kingdom

B

M & S, hello from Syria. Just had a week in a Bedouin camp. Tilly did cooking and weaving, and I'm a camel expert now!

Most nights we slept out under the stars. I've never seen so many stars. Incredible.

I think I'll just stay here. Say goodbye to the boss for me.
Tom

Mike and Seetha
4100 W Wilson Ave
Chicago
IL 60630
USA

C

Hi all

Taj Mahal stunning, and such a romantic story. What's missing from this picture is thousands of tourists (including us).

Kerala next stop, and a 4-day cruise on a houseboat. Eat your hearts out.
xxx
Chris, Kylie, Sam

The inmates
16 Carabella Street
Kirribilli
Syndey
NSW
Australia 2061

So, what's a tourist?

A tourist is a person who travels to a place that is not his or her usual place, and stays there for at least a night. It could be on holiday, or for business, or another purpose.

But this chapter is mainly about people on holiday.

Did you know?
- You can already book space travel!
- Several tourists have been to the International Space Station (350 km up).

What's tourism?

Tourism means all the activities that tourists take part in, and the services that support them. Hotels, airports, taxi drivers, and ice cream sellers at the seaside: all are part of the tourism industry.

Holiday tourism: the essentials

So what does a place need, for holiday tourism? These are the essentials:

1 An attraction You need an attraction of some kind to bring the tourists in. It could be a natural attraction, or a built one, or a mix. It could even just be peace and quiet!

2 Accommodation and catering Tourists need places to sleep and wash. And they need food and drink.

3 Transport There must be a way for the tourists to reach the attraction – by road, or footpath, or air, or sea.

If any of these essentials is missing, or not good enough, or if the attraction gets spoiled, you won't have much of a tourism industry.

▲ *Tourists need attractions. And accommodation, catering, transport …*

Your turn

1 Look at postcards **A – C**. In which photo would you say the main attraction is:
  - **a** the local people's way of life?
  - **b** a built feature?
  - **c** a natural feature?

2 Most countries have some natural tourist attractions.
  - **a** Make a copy of the spider map started below. (You will add a lot to it, so draw it in the middle of a new page.)

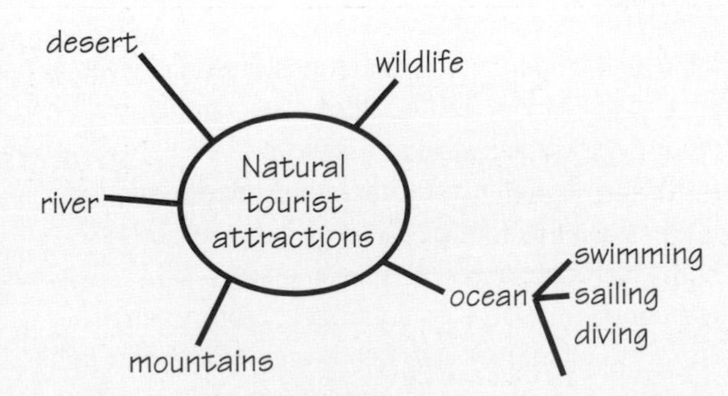

  - **b** Add any other natural attractions you can think of. (Could *climate* be one?)
  - **c** Now write in activities for tourists, based on these features. (This has been started, for the ocean.)

3 Many tourist attractions are *built*. Some are built just for tourists. Disney World is an example.
  - **a** List as many built tourist attractions as you can.
  - **b** After each one, write its country in brackets.

4 Tourists need accommodation and catering too. Copy this spider map and add as many more examples as you can:

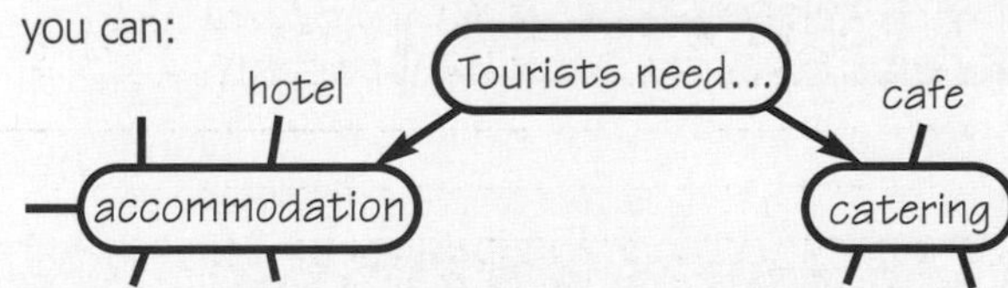

5 The final ingredient for the tourism business is transport. Make up your own way of showing the different options for this. (Cars, coaches, …)

6 **a** Using what you've learned on this page, list all the jobs you can think of, connected to tourism.
  - **b** Underline any from the primary sector in one colour, the secondary in another, and the tertiary in a third. Add a colour key. What do you notice?

7 Look again at postcards A – C. Do you think any of those three places could be spoiled by tourism? For any that you choose, explain why.

World tourism

This unit is about tourism as a way for countries to earn money.

The tourism business

Tourism is fun ... and a deadly serious business. Look at this:

- In 2007, the world spent about 3 billion US dollars *a day* on **international tourism** (where you travel to other countries). Most of it was for holidays.
- For many poorer countries, tourism is one of their few ways to earn money.
- In many countries, people also spend a lot on **domestic tourism**, where you are a tourist in your own country.
- 1 in every 12 jobs around the world is linked to tourism.

▲ *A big welcome for tourists – and the money they bring in.*

Now off you go to explore some data.

Your turn

1 Look at graph **A**.
 a What do you think *international tourist arrivals* means?
 b About how many of these were there in 1960?
 c About how many were there in 2005?

2 Here's one reason for the growth in world tourism:
In many countries, incomes have risen quite a lot.
See how many other reasons you can think of.
For example, might TV play a part?

3 Table **B** shows the top destination countries, for 2007.
 a In which continent are most of these?
 b France had top place. Can you suggest any reasons for this? (For example, how might its location help?)
 c How many *fewer* tourists did the UK get than France, that year? Try to give reasons for this difference.

4 Look at graph **C**.
 a **i** Which two months had most arrivals?
 ii See if you can explain this.
 b Which month had fewest arrivals?
 c There's a little peak for December. Why?
 d What problems do you think this pattern of tourist arrivals could cause for:
 i tourist destinations? **ii** people working in tourism?

5 Now look at graph **D**. It shows how tourism grew from 1950 to 2007, with forecasts from then until 2020.
 a Which region had *fewest* international tourists, in 2007? See if you can suggest a reason for this.
 b Overall, is world tourism expected to *grow*, or *shrink*, between now and 2020?
 c In which region is tourism expected to grow *fastest*, between now and 2020? How did you decide?

6 In **D**, the tourist numbers from 2007 to 2020 are just forecasts. Try to say how each of these might affect the *actual* numbers.

 a Oil prices rise fast, as oil begins to run out.
 b China grows wealthy – much faster than everyone expected.
 c People get very worried about CO_2 emissions.
 d Global warming causes rapid climate change.

7 In table **E**, the money from inbound tourists is given as a % of all the money earned from other countries.
 a What do you think *inbound tourists* are?
 b **i** In which country in the table are people best off?
 ii What % of its foreign earnings is from tourism?
 c **i** In which country are the people worst off?
 ii What % of its foreign earnings is from tourism?
 d From the table, pick out the country where you think people might suffer most, if tourism declined. Explain your choice.

8 So far, do you think tourism is a good thing? Give your answer in 35 words.

World tourism: some data

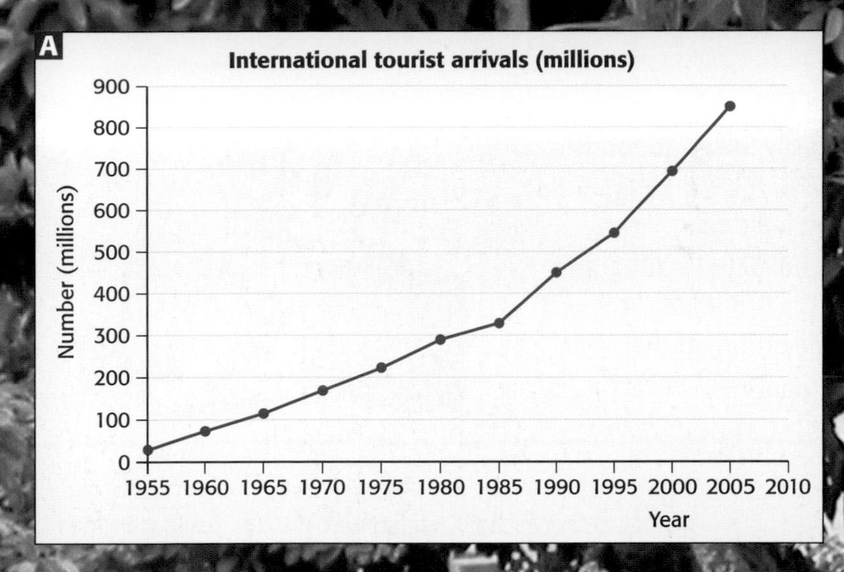

B Top ten destinations for international tourists, in 2007

Rank	Country	Tourists (millions)
1	France	81.9
2	Spain	59.2
3	USA	56.0
4	China	54.7
5	Italy	43.7
6	UK	30.7
7	Germany	24.4
8	Ukraine	23.1
9	Turkey	22.2
10	Mexico	21.4

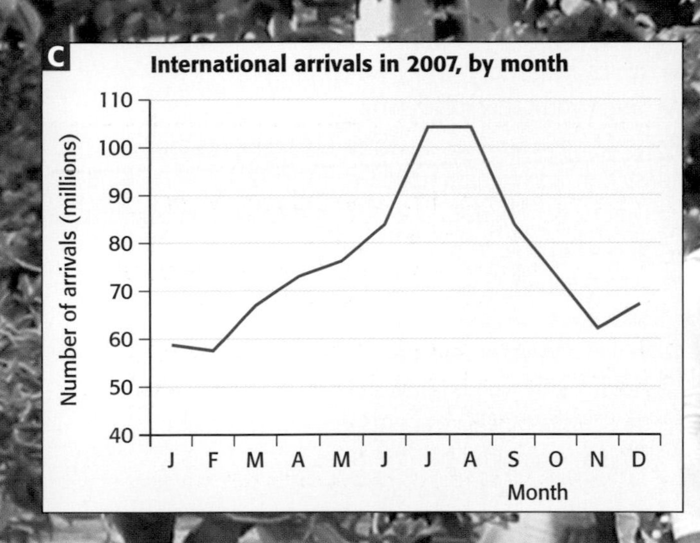

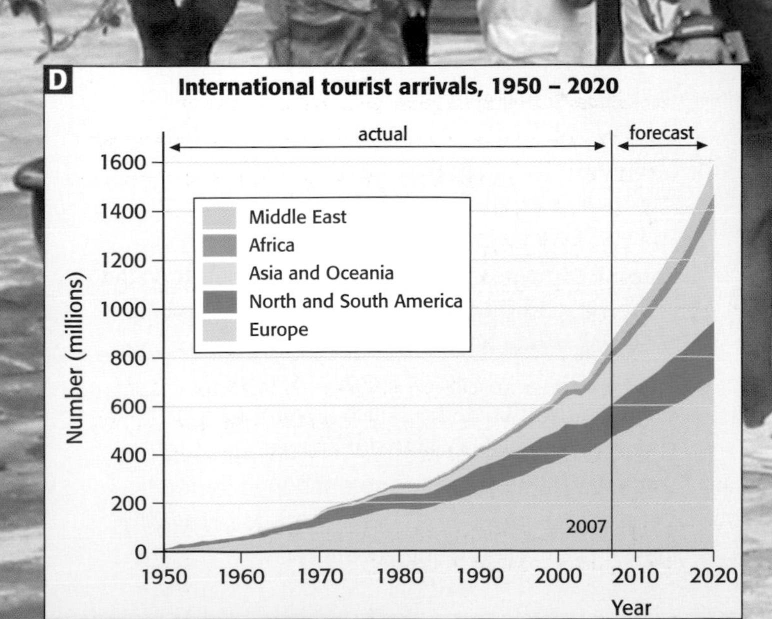

E

Country	GDP per capita (US$ PPP)	Money from inbound tourists as a % of foreign earnings
Barbados	17 300	48%
Brazil	8400	6%
Cambodia	2700	23%
China	6800	7%
Egypt	4300	27%
France	30 400	18%
Gambia	1900	22%
Saint Lucia	6700	57%
Seychelles	16 100	41%
UK	33 200	11%

Tourism in the UK

Here you'll think about the UK's tourist attractions, and explore its holiday patterns.

Tourism in the UK

Tourism is very important to the UK. Look at these facts:

- 1.45 million people in the UK work in jobs directly linked to tourism.
- In 2007, around 11 million holidays were taken in the UK by inbound tourists. They spent around £5.4 billion on their holidays.
- UK residents took around 77 million domestic holidays in 2007, and spent about £14.0 billion.
- UK residents also took around 45 million holidays abroad that year.

Now it's time to be a data detective, and look for patterns.

Your turn

First, the UK's tourist attractions

1 The UK has lots of tourist attractions. See how many examples you can give, for each of these:
- **a** built attractions (the British Museum, Alton Towers ...)
- **b** seaside resorts (Brighton, ...)
- **c** historic sites (Stonehenge, ...)
- **d** National Parks (the Lake District, ...)
- **e** buildings linked to the government, and Royal Family
- **f** sites linked to sports, and well-known teams
- **g** other places linked to famous people

Where UK residents take their holidays

2 This question is about graph **A** on page 105.
- **a** One line shows *domestic* holidays. Which one?
- **b** From the graph, what was the *overall* trend for:
  - **i** total holidays taken? (Did the number fall?)
  - **ii** domestic holidays taken?
  - **iii** foreign holidays taken?
- **c** In which year did people first take more holidays abroad (of 4+ nights) than at home?

3 Now, for **A**, see if you can suggest reasons why:
- **a** the total number of holidays grew, in that period
- **b** the number of foreign holidays increased
- **c** the *total* line is not smooth, but zig zags.

There are clues in the boxes below. But use what you learned in Unit 6.2 too, and your own knowledge.

Tour operators compete with each other.

Jumbo jets were introduced in 1971. They carry more people, further, and more cheaply.

The economy is in better shape some years than others.

TV?

Weather?

Households in the UK with use of a car: 31% in 1971, 72% in 2000.

The average wage in the UK has risen steadily since 1971.

Airlines like easyJet and Ryanair?

4 Look at **B**. Which UK region was the top destination for domestic holidays that year? See if you can name any of the counties in this region, and tourist attractions.

5 Table **C** shows the length of those domestic holidays.
- **a** Which length of holiday was more popular?
- **b** See if you can explain why.

6 Table **D** lists countries in which UK residents took at least 1 million foreign holidays, in 2007.
- **a** Draw a bar chart for the data in the first two columns. Show the bars in order of height.
- **b** Which country was by far the most popular? See if you can explain why.
- **c** Which *continent* was the most popular? Why?
- **d** Now see if you can pick out two countries where most visits by UK residents were short breaks. (The third column has the clues.)

Inbound tourists

7 Table **E** shows the top ten countries that our inbound tourists came from, in 2007.
- **a** Which were the top three, by number of tourists?
- **b** Which *continents* are represented in the table?

8 Overall, which holiday tourists are more important to the British economy: *inbound* tourists or *domestic* tourists? Give your evidence.

9 Graph **F** compares how much inbound tourists spend in the UK, and UK residents spend on trips abroad.
- **a** Overall, which group spends more?
- **b** Is the spending gap *growing*, or *shrinking*?
- **c** How do you think the British government might feel about this? Explain your answer.

10 Your job is to attract even more inbound tourists! Design some pages (1–3) for a website saying why the UK is a really brilliant place to visit.

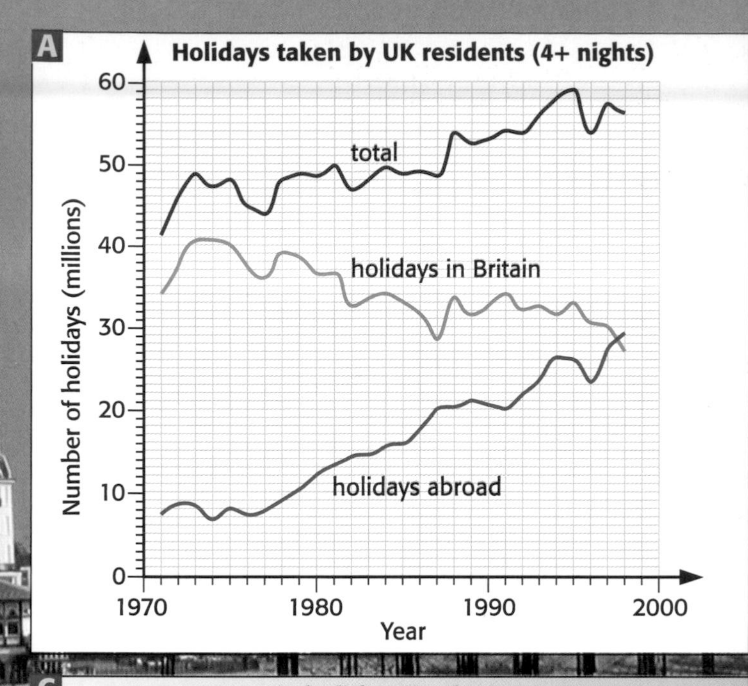

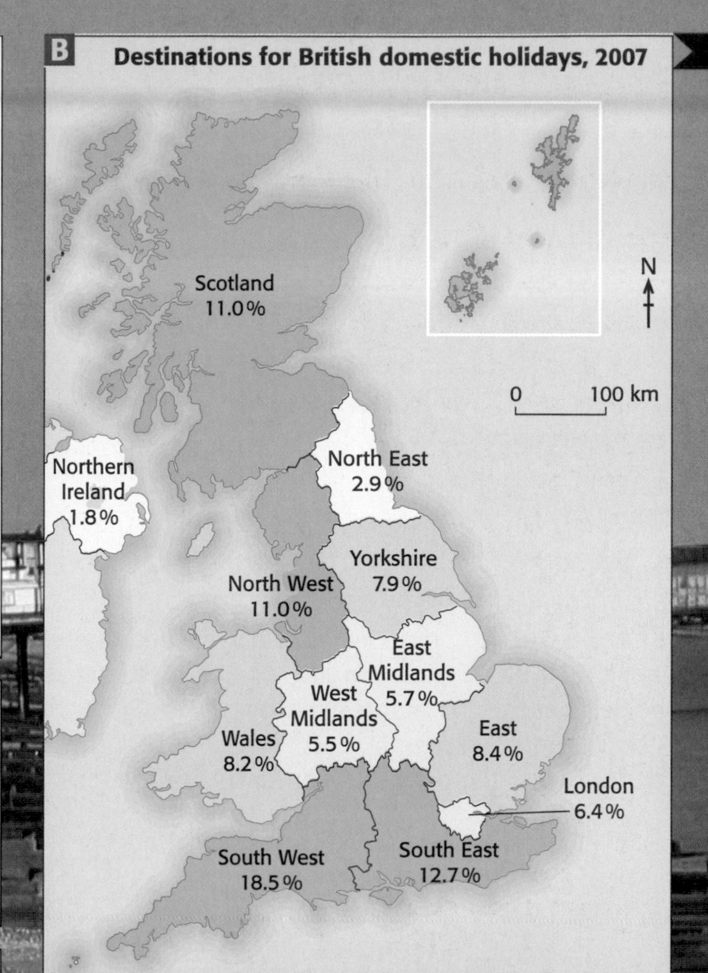

C Domestic holidays in the UK, 2007

Length	Number taken (millions)	Amount spent (£ billions)
1 – 3 nights	49.54	6.9
4+ nights	27.28	7.1
Total	76.82	14.0

D Top holiday destinations abroad, for UK residents in 2007

Destination	Visits by UK residents (millions)	Average length of stay (nights)
Belgium	1.0	3
Cyprus	1.1	11
France	7.6	7
Greece	2.3	10
Irish Republic	1.5	6
Italy	2.6	7
Netherlands	1.1	4
Spain	12.0	10
Turkey	1.3	10
USA	2.4	14
rest of world	12.5	

E Top 10 source countries for inbound tourists taking holidays in the UK, 2007

Country	Holiday visits to the UK from it (millions)	Amount spent (£ billions)
Australia	0.42	0.29
Canada	0.28	0.19
Belgium	0.37	0.09
France	1.17	0.28
Germany	1.18	0.46
Irish Republic	0.70	0.24
Italy	0.65	0.30
Netherlands	0.62	0.20
Spain	0.75	0.32
USA	1.36	0.91
rest of world	3.29	2.07
World total	**10.79**	**5.35**

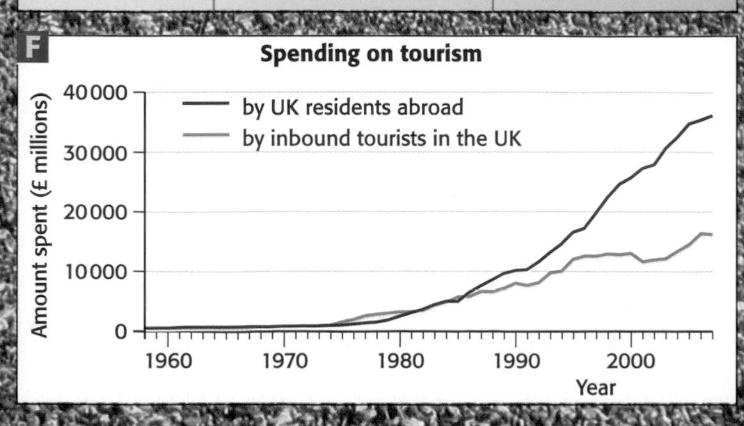

Now imagine …

This unit tells you a story, to introduce the pros and cons of tourism.

Your paradise home

Imagine you live on an island. Let's call it Xi. It has a great climate, golden beaches, thick forests, exotic wildlife, and 100 people. You farm and fish. You are poor, but content.

Then one day a boat arrives, with 10 strangers. To see the island! You give them food and a place to sleep. They give you money. In a few weeks they go off again, happy.

Some islanders decide to put up new huts, in case more strangers arrive. And next year, 20 turn up! They go everywhere, but mainly onto your beaches. That's okay.

The years go by. Each year, more huts get built. More strangers arrive. They bring more money. For a few weeks they relax and have fun. And then they leave.

The money is helping the island. You have built a school hut, and a clinic hut. You have hired a teacher and nurse from the mainland. All the children go to school. Great!

But some islanders, and even some strangers, get greedy. They take over land that was once shared. They build huge huts. They invite lots more strangers. Not so good!

Now there's no room for you on your own beaches. More and more trees are being cut down, for new huts. You notice there's a lot less wildlife – and a lot more rubbish!

You begin to resent the crowds of strangers everywhere. And their skimpy clothes, and bad manners. You depend on their money now – but you can't wait for them to go.

Most of the islanders feel like you. They are no longer content. There is conflict everywhere. You feel you have gained some things, but lost a lot. Where did it all go wrong?

Just a story?

Xi was an imaginary island.

Tourist hotspots are not just little islands, with golden beaches. They can be pretty villages, historic towns and cities, ski resorts, seaside resorts, ancient sites, National Parks, and more.

Tourists enjoy all of these. And all may benefit from tourism. But if it is not managed properly, things can go very wrong.

▲ Here, Filipino farmers protest about farmland being taken over for tourism.

▲ Here, Venetians protest about plans for more tourist accommodation. The banner reads 'Venice is not an hotel'.

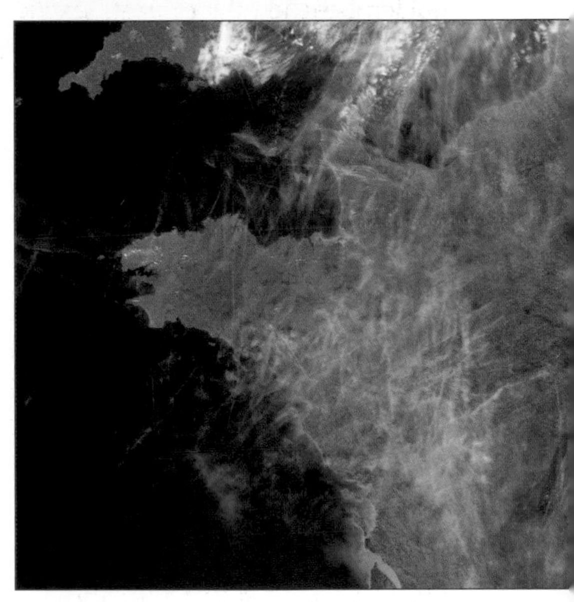

▲ *And there go the tourists! The white trails are water vapour from plane engines, which turns to ice. The engines give out carbon dioxide too. (Why?)*

Sustainable tourism

Now there is growing pressure to make tourism sustainable. **Sustainable tourism** means:

- a place, its people, and their culture, are respected
- the local people have a say in the decisions about tourism
- they gain a fair share of the benefits from it, including money
- there is as little damage to the environment as possible.

In the next three units, we look at examples of tourism in different places. As you go through these, ask yourself whether the tourism is sustainable.

Your turn

1 From the story on page 106, list any benefits of tourism you can identify:
  a for the tourists **b** for the people of Xi

2 Now list any negative consequences of tourism you can identify, for Xi and its people.

3 Look at your list in **2**.
  a First, underline any consequences you think could have been avoided.
  b For each one you underlined, suggest how it could have been avoided.

4 **a** What does the word *sustainable* mean? (Glossary?)
  b In 20 words of your own, explain what *sustainable tourism* is.
  c Do you think the tourism on Xi was sustainable? Give some reasons for your answer.

5 Most tourism is by plane, car, and coach. So do you think it can ever be *truly* sustainable? Explain.

6 The story of Xi has some lessons for tourists too. See if you can think up *Three Golden Rules for Tourists*, with help from the story.

6.5 Beautiful Benidorm?

SPAIN
Benidorm

Here you will see how mass tourism can change a place completely, and forever.

Look at the change ...

Benidorm, in Spain, in 1960: 6200 residents, two great beaches, and just a few hotels.

Benidorm today: 68 000 residents, and over 550 000 tourists at peak season. (Most of them British!)

... thanks to the package holiday

In 1950, very few people in the UK went abroad on holiday. But wages were rising. War planes from World War II were being converted to carry passengers. A group of businessmen toured the Mediterranean, looking for good holiday spots ... and the **package holiday** was born.

In 1957 the first package tour, from the UK, arrived in Benidorm. The tourists loved its sunshine, peace and quiet.

News spread. And soon new hotels and apartments were springing up everywhere, to meet demand. But there was little planning or control. Much of the building was poor quality.

How the package holiday works

- A tour operator selects a hotel.
- It books a block of rooms for next season (or several seasons).
- It also books some planes (or may even buy its own).
- Then it sells a complete holiday (flight + hotel and at least some meals) to tourists.

Going downhill

25 years later, Benidorm was not so great. It was showing the downside of **mass tourism**.

- People were packed on the beaches like sardines.
- Hotels had been built tall, to allow open space between them – but the result was a forest of concrete.
- Because it was cheap it attracted lots of 'lager louts'.
- The crowds, and loud discos, and noisy nightlife, put families off.
- Many of the hotels were shoddy.
- There was little trace of the real Spain. It was easier to get fish and chips than Spanish food.

So tourists began to stay away ...

Doing better now

Benidorm is very important to Spain: it produces 1% of its GDP. So the government got worried about it, and began to take more control.

Now, poor hotels have been improved. New upmarket ones have been built. Plus a new theme park close by, to attract visitors all year.

Benidorm says it is **eco-friendly** now, and a model for mass tourism. Why? Because it caters well for lots of visitors, in a small space. (4 million a year!) It is kept really clean. The hotel appliances save energy and water. Much of the food is from the local farmers. And you can walk everywhere.

One problem …

Benidorm uses huge amounts of water, for tourist swimming pools and showers. This diagram shows what's happening:

Did you know?
- The Benidorm area has around 30 000 swimming pools!

Benidorm has been fixing all its leaky pipes, to save water. The new theme park uses sea water in its lakes, and recycled water in its gardens. But Spain is a dry country. The water problem will not go away.

And it's not just Benidorm. Water supply is a problem in many tourist resorts around the world. Tourists are water-hungry!

Your turn

1 What do these terms from the text mean? (Glossary?)
 a package holiday b mass tourism c eco-friendly

2 a What were the attractions of Benidorm in 1960?
 b Say how mass tourism has affected each of them.

3 a Make a *large* copy of the 'vicious circle' below.
 b Then write these sentences in the boxes in the right order, to show how tourism can ruin a place. Don't give their labels. (Hint: put sentence **C** in box ①.)
 A In the end, no tourists want to go there at all.
 B So developers rush to build new tourist facilities.
 C Tour operators offer cheap packages to a resort.
 D But development isn't managed or controlled …
 E Now many tourists are put off.
 F Tourists rush to book because it's so cheap.
 G … so the resort's natural attractions get ruined.
 H So the tour operators have to slash prices further.

4 Look at your vicious circle in **3**.
 a A government *could* object at step ①. Give a reason why it might not wish to.
 b At what point did the Spanish government intervene in Benidorm's development?
 c At what step do you think it *should* have done so?

5 If tourism ruins a place, whose fault is it?
 Using your vicious circle to help you, list all the groups you think may be to blame. And give your reasons.

6 Benidorm gets lots of sun. Come up with a sustainable way to give it as much clean water as it needs. (Hints: solar, sea.) Include a drawing of your scheme.

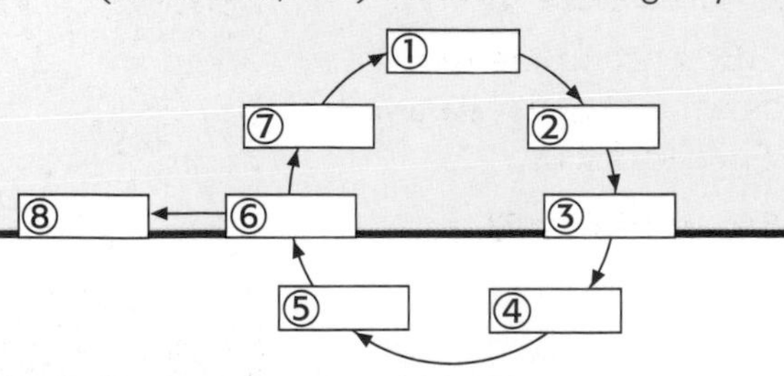

How much does Gambia gain?

Tourism can help poorer countries – but perhaps not as much as you'd expect. We take Gambia as example.

About Gambia

- Gambia is a small country of about 1.7 million people, in West Africa. It was a British colony until 1965. It is 90% Muslim.
- It is very poor. It has no fossil fuels or metal ores. Its main export is peanuts. 75% of Gambians depend on farming.
- But it does have some big attractions: golden beaches, friendly people, and a warm, dry, and sunny climate during our winter months.
- So, like many LEDCs, Gambia relies on tourism as a way to earn money. It has over 110 000 tourists a year. (Over half are from the UK.) About 18% of its total GDP is from tourism.

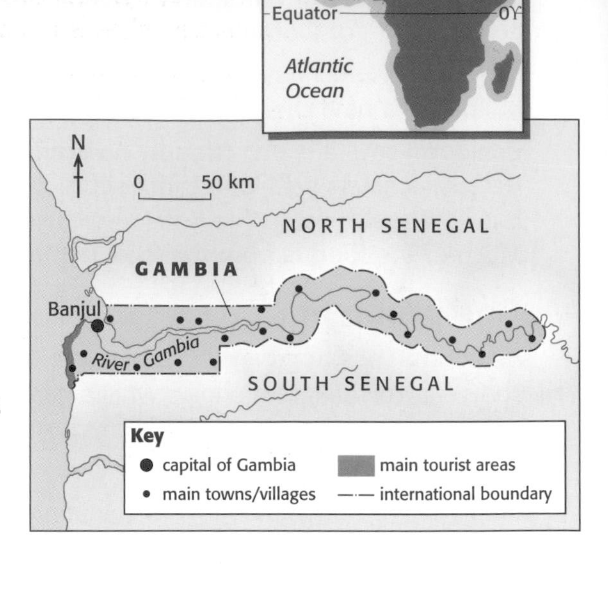

So is everyone happy?

Overall, Gambians welcome tourists. They need their money. Around 1 in 7 of Gambia's workers are in jobs related to tourism. Without tourists, Gambia would be poorer.

But tourism has also brought some conflicts. Like these ...

A very leaky business

Gambians are right when they say that Gambia does not see so much of the money the tourists pay. It leaks out all over the place! Like this …

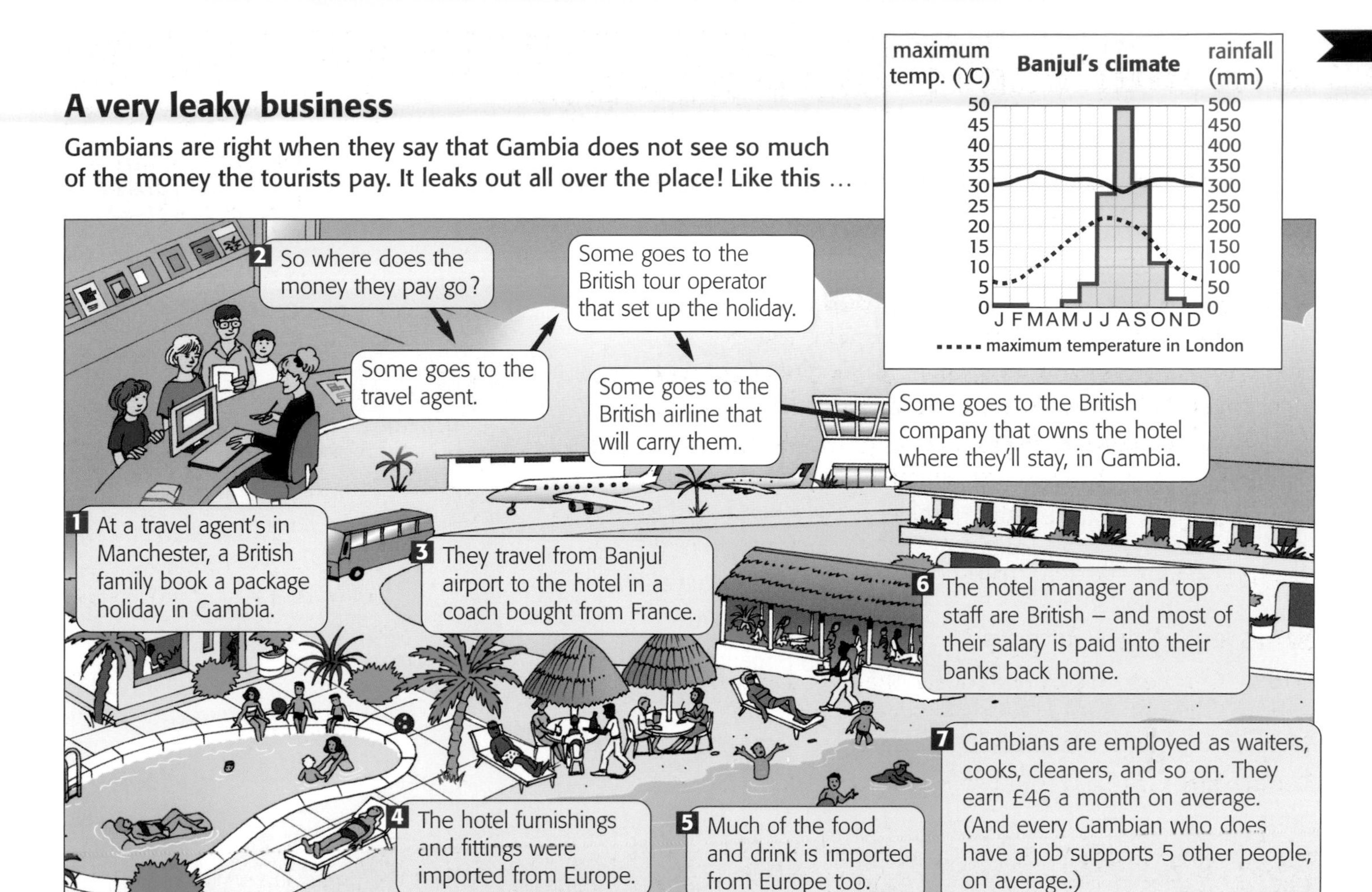

So Gambia 'owns' the natural attractions. But overall, it gets less than one-third of what the tourists spend. Do you think that's fair?

Your turn

1 Using the map on page 110, write a paragraph on the geography of Gambia: where it is, physical features, roughly how long and wide it is, and so on.

2 One of Gambia's main attractions is its climate. Look at the climate graph at the top of this page.
  a Which two months are likely to bring most British tourists? Why?
  b In which three months are hotel staff most likely to be laid off? Why?

3 Look at this table comparing Gambia and the UK.

	GDP per capita (US$ PPP)	doctors per 100 000 people	% under-nourished
Gambia	1920	11	29
UK	33 240	230	very low

  a From the table, what can you conclude about:
    i the level of poverty in Gambia?
    ii the level of development in Gambia?
  b Gambians think that British tourists are wealthy. Do they have good reason for this?

4 Tourism could help Gambia to develop. Draw a consequence map, like the one started here, to explain why.

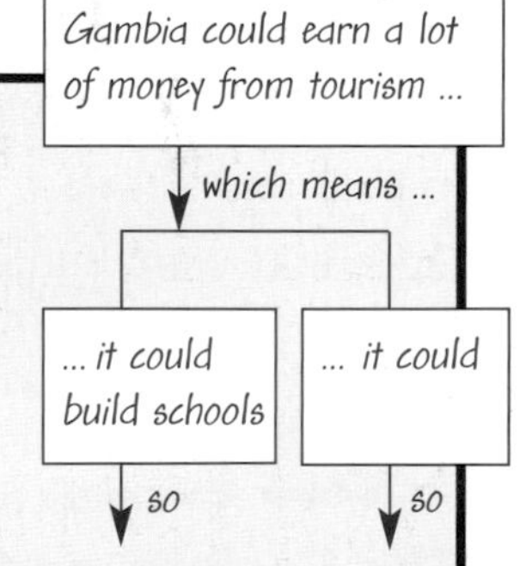

5 But tourism in Gambia benefits other countries more than Gambia.
  a Why is this?
  b So some tourists try to shop in local shops, and eat in local restaurants. Some even stay in people's homes. How does that help Gambia?

6 What do you think this person would give as the main:
  a advantage **b** disadvantage of tourism?

- A the prime minister of Gambia
- B a Gambian farmer
- C a waiter in a tourist hotel in Gambia
- D a strict Gambian mum, with three teenage sons

7 You are Gambia's Minister of Tourism. You want Gambia to gain more from tourism. What steps will you take? Write your answer as a speech to Gambia's parliament.

Ecotourism among the Ese'eja

Here you'll find out what ecotourism is. We go to the rainforest in Peru for our example.

Did you know?
- In many LEDCs, local people take second place to tourists.
- Does that happen in the UK?

What is ecotourism?

In Unit 6.5 you saw that Benidorm is a destination for mass tourism. **Ecotourism** is very different, as these panels show.

Who are the Ese'eja?

The Ese'eja are a rainforest tribe. They live in the Amazon rainforest in Peru and Bolivia.

Once they were hunter gatherers, moving around in search of food. Now they live a more settled life. They grow crops, and keep cattle. (For this they clear some rainforest.) They still hunt a bit, and fish, and gather brazil nuts from the forest to sell.

The Ese'eja and ecotourism

Some Ese'eja live in a **reserve** in the rainforest in south-east Peru. They number about 400. And they're in the ecotourism business!

In 1996 they signed a 20-year contract with a Peruvian tour operator. They agreed to build a tourist lodge, in the style of their own homes. The tour operator agreed to bring tourists along. And here's the deal:

- the Ese'eja get 60% of the profits from the lodge
- they have an equal say with the tour operator, in all the decisions
- they protect animals they once hunted, so tourists can see them
- they are trained to do most of the jobs in the lodge
- they act as rainforest guides for the tourists
- they sell food to the lodge
- they sell tourists the woven baskets, fans, and other items they make.

Hey Jude

Back in Lima again, after 3 days at the Posada Amazonas. (That's the Ese'eja tourist lodge I was telling you about.)

It was different! Our bedroom was open on one side to the rainforest, and lit at night by candlelight. So we drifted off to sleep to a great chorus of squeaks and croaks and chirps.

Once something flew around the room in the middle of the night. I think it was a bat. And things kept dropping from the ceiling. But I felt safe in my mosquito net.

One morning, really early, we climbed up a 35-metre tower, to the top of the rainforest. We saw parrots, and macaws, and toucans. We had a trip on a catamaran too, and saw otters, and stork, and monkeys, and capybara grazing on the bank.

But what I liked best were the rainforest walks in the dark, with torches. You had to step carefully. But you could see little eyes glowing everywhere. Frogs, and all kinds of insects and beetles, and even a few snakes. The guides were brilliant. They answered all our questions, and explained everything really well.

Home next week. Too bad – I love it here. See you soon.
Hugs

Sally

▲ *Your room awaits at the Posada Amazonas (the Ese'eja tourist lodge).*

▲ *You might see capybara – like giant guinea pigs, up to 130 cm long.*

▲ *At the top of the rainforest tower.*

The benefits of the project

- It brings in money for the Ese'eja people, who are very poor.
- They are using their skills in farming and fishing, and their knowledge of the rainforest.
- They are also learning new skills (including speaking English).
- The local wildlife has benefited too. The Ese'eja protect it because it is part of the tourist attraction.

Your turn

1 The Ese'eja live in a reserve in Peru's rainforest.
  a What is a *reserve*?
  b Where is Peru? Which countries does it border?

2 a What does *sustainable* mean?
  b Do you think the Ese'eja ecotourism project is an example of sustainable tourism?
  c Give it a mark out of 10 for sustainability, and explain why it deserves that score.

3 The tourists who stay at Posada Amazonas (the lodge) are *ecotourists*. What is an ecotourist?

4 Suppose lots of people hear about Posada Amazonas, and start turning up in their hundreds. Would that be sustainable tourism? Explain.

5 Do you think projects like Posada Amazonas could solve some of Gambia's tourism problems? Explain your answer.

6 But not everyone thinks ecotourism is a good thing. Do you agree with this person? Give your response, with reasons.

If you really want to protect a place and its people ... don't go there!

A challenge in the Broads

Here you'll learn about the Broads National Park, and design a sustainable activity centre there.

The Broads

The UK has 14 National Parks. These are large areas of special beauty, protected by law for us all to enjoy.

The **Norfolk and Suffolk Broads** is one. This low flat area is a web of shallow lakes, rivers, marshes, and dykes. Around 5700 people live here, most in small villages. And around 1 million more visit each year!

It's not natural

The lake-filled landscape may look natural– but it's not. For centuries, people dug up the local peat for fuel, leaving wide, shallow pits. Over time, these filled with water ... and today's landscape began to take shape.

The Norfolk and Suffolk Broads

Want a relaxing boating holiday?
Want to learn to kayak, or windsurf?
Or take up painting?
Or go birdwatching, or fishing?
Watch the world sail by, from your riverbank?
Bike for miles, with never a hill?
Visit windmills, and explore charming villages?
Then the Broads is the place for you.

We offer over 200 km of waterways. Rare wildlife. Sailing. Boats and bikes to hire. Cycle paths. Walks. A calm refuge in a hectic world. Come visit.

A

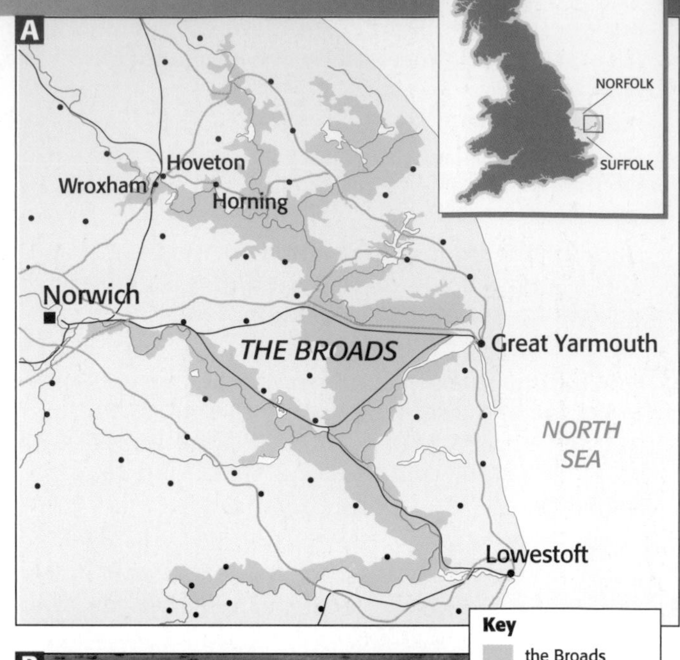

B

▲ *Sailing in the Broads.*

C

D

▲ *The bustling little town of Wroxham is called the capital of the Broads. It has a population of around 1500.*

◀ *Looking down over Wroxham Broad. Find it on the OS map on the next page.*

Your turn

1 a Where in the UK are the Broads?
b What's the landscape like there? (Check the photos.)

2 Look at the OS map above. It shows just a small part of the Broads, including Wroxham.
a The word *Broad* appears several times on the map. What do you think it means?
b How can you tell from the map that the land is flat?
c Look for the highest point you can find, on the map. Give a grid reference. How high above sea level is it?

3 What evidence can you find from the map:
a that this may be a good place to observe birdife?
b that there is plenty for tourists to do? (Page 138?)

4 a Compare photo **C** and the OS map. In which direction was the camera pointing? (Hint: the Yacht Club ...)
b One field is marked **X** on the photo. See if you can find this field on the OS map. Give a grid reference.

5 Now, a challenge for you. The owners of field **X** want to build an activity centre on it, for young people. And they'd like you to design it! Here are their notes:

- *The centre should have access to the water.*
- *It must offer a wide range of activities, with tutors.*
- *It can be residential, if you wish.*
- *Keep sustainability in mind. For example, how will the centre benefit local people? How will people travel to it?*

a First, plan your centre. Will it be residential? How many young people at a time? What activities? How will people get there? (Check out Hoveton!) What else will you do, to promote sustainability?
b Now write up your proposal. Try to include sketches, and rough plans. Photo **C** and the OS map will help.

7 The ocean

The big picture

This chapter is about the ocean. These are the big ideas behind the chapter:

- 71% of our planet is covered by ocean. Or nearly three-quarters!
- We can't live without the ocean. It plays an enormous part in the natural life of the planet.
- We humans make use of the ocean in many different ways.
- We are harming the ocean ecosystem.

Your goals for this chapter

By the end of this chapter you should be able to answer these questions:

- The global ocean is divided into smaller oceans. How many? What are their names? And where are they?
- What do these terms mean?
  ocean ridge *ocean trench* *continental shelf* *coral reef*
- The ocean has high ridges, deep trenches, and lots of volcanoes. Explain why.
- What do these terms mean?
  phytoplankton *water cycle* *ocean current* *global conveyor*
- The ocean plays a major part in the natural life of our planet. How? (Describe the four main ways.)
- We make much use of the ocean. In what ways? (Give at least six.)
- We are harming the ocean ecosystem. In what ways? (Describe all three main ways.)

And then ...

When you finish the chapter, come back to this page and see if you have met your goals!

Did you know?

- There are many more volcanoes under the ocean than on land.

Did you know?

- The largest animal on the planet is the Blue Whale.
- It can grow to over 30 m long.

Did you know?

- The UK has its own coral reefs.
- They are cold-water reefs, and lie north of Scotland.

What if ...

- ... we woke up one day to find the ocean had gone?

Your chapter starter

Look at the photo on page 116. Where do you think that place is?

How much can you say about the brightly-coloured animal?

Why do you think the humans are there?

What steps have they taken, to adapt to this environment?

Look at those things at the bottom. Do you think they are plants?

Our watery planet

Nearly three-quarters of the Earth is covered by ocean. Find out more here.

The Blue Planet

Our Earth is sometimes called the Blue Planet, because it looks blue from space. That's because 71% of its surface is covered by **ocean**!

- The ocean covers about 360 million sq km of the Earth.
- In nearly half of it, the water is over 3 km deep. Think about that! And in the very deepest part, it is almost 11 km deep.
- The water is **salt water**. It contains sodium chloride and other salts. If you drink it, they make you dehydrated. You could die.
- By contrast, the water in rivers is not salty. It is called **fresh water**.

We can't live in the ocean. A lot of the Earth's land does not suit us either. In fact, only about 13% of the Earth's surface is habitable, for us humans. Does this make you feel squashed?

▲ *The Blue Planet, your home – 71% covered by ocean.*

A map of the ocean

We have divided the global ocean into smaller ones. Look at this map. Note that they are all joined, and water flows between them.

- The Pacific Ocean is the largest.
- The Arctic Ocean is the smallest, and it's covered in ice in winter.
- We often give parts of an ocean special names, at the coast. Like **bay**, **sea**, **gulf**, and **channel.** They are still part of the ocean! Some are named here, in red.

Now have a look at the map and information on the next page.

Then try 'Your turn'.

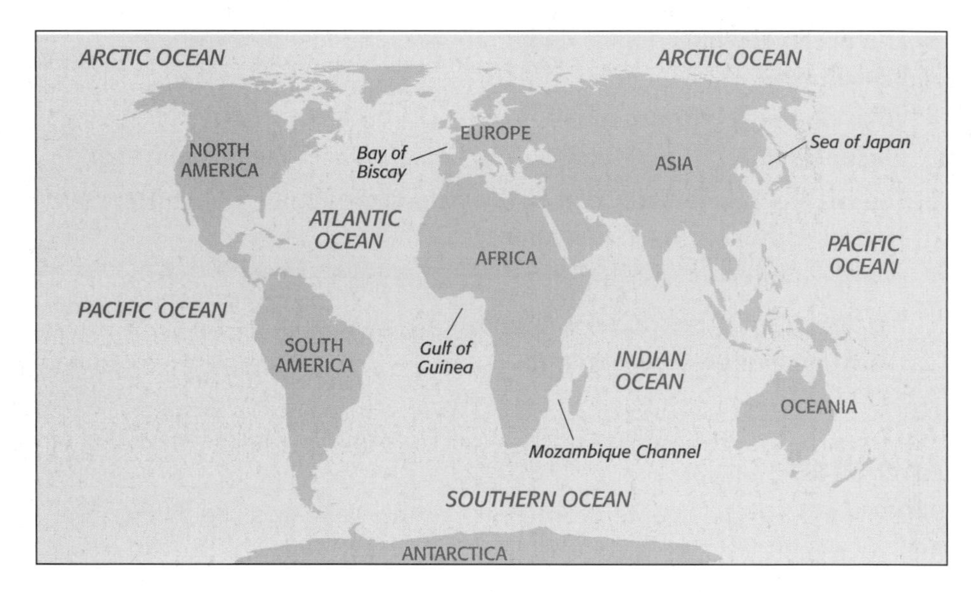

Your turn

1 Name the ocean that connects:
a the UK and West Africa **b** Kenya and India
c Japan and the USA **d** South Africa and Antarctica

2 See if you can suggest a sensible route for a cargo ship:
a from China to Alaska **b** from China to Ghana
Each time, name the ocean(s) the ship will pass through, and countries it will pass by. (Pages 140 – 141?)

3 Look at the map on page 119. What can you say about the depth of the ocean:
a at A? **b** at B? **c** around the UK?

4 Name the feature on the map at: **i** D **ii** C **iii** B

5 Now write out this paragraph – but with the jumbled words unjumbled. (Use what you know already!)

> The ocean floor has staminnou and deep chertnes. They result from latep movements. So you can expect quathseekar too. These can cause giant waves called imutsna, that travel across the ocean.

6 Shock! Horror! One night, aliens from another planet siphon the Earth's ocean away.
You are sent to report on the empty ocean basin. Describe what you see, and feel, and smell, in the form of a blog or radio report. Make it gripping!

The ocean floor

Imagine you can drop down to the ocean floor, in a special pod, to explore it. What will you find there?

1 You'll find large flat areas covered with a layer of muddy sediment – nearly half a kilometre thick, on average. And below it is rock.

On and in the sediment, you'll find many living things. And dead things.

2 You will also find long mountain ranges! They are called **ridges**, and show where the Earth's plates are moving apart.

They are built up by magma, rising from the Earth's mantle:

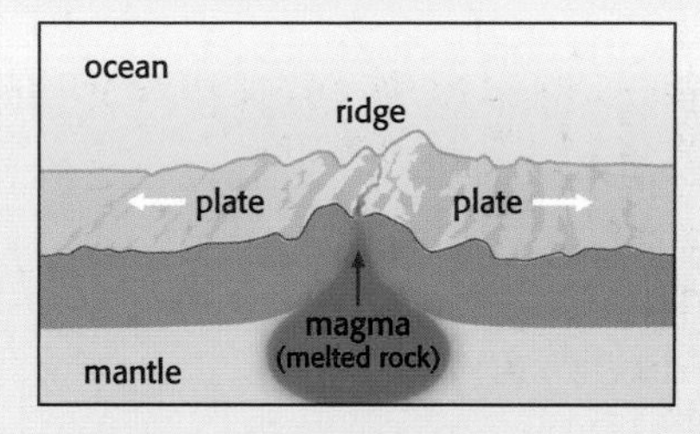

The Mid-Atlantic Ridge, below, is the longest mountain range on our planet!

3 And you will find deep trenches, where one plate is being forced under another.

The deepest one is the **Mariana Trench**, near Japan. It plunges to 10 920 m below sea level.

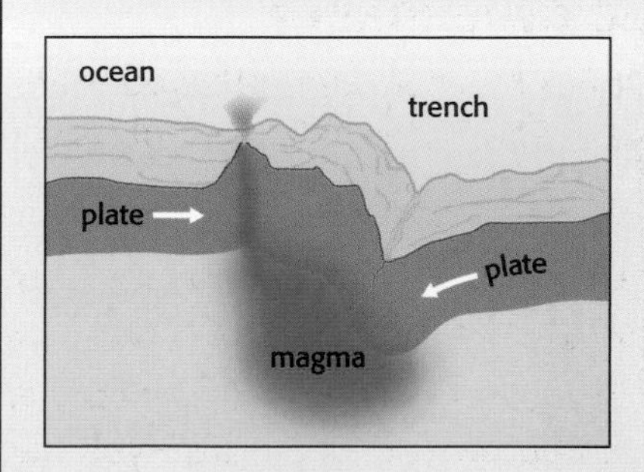

Depth of ocean floor below sea level (m)

0
200
500
1000
2000
3000
4000
5000
6000
7000

A
B
C
D

4 You'll find thousands of volcanoes – along ocean ridges, beside ocean trenches, and at other 'hotspots' where magma erupts through the ocean floor.

If you're lucky, you might see one erupting, like here!

5 Look at the paler areas along most of the coasts. They show where the land slopes gently into the ocean, forming a **continental shelf**:

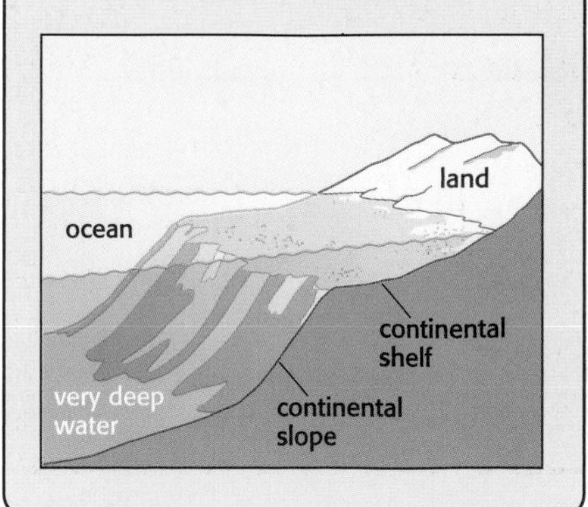

6 In warm shallow ocean water, you may come across **coral reefs**.

These are built up of limestone, secreted by animals called **polyps**.

They are home to an amazing variety of fish, sponges, anemones, and other animals.

Australia's **Great Barrier Reef** is the largest reef system in the world.

The essential ocean

This unit is about the major role the ocean plays, in the life of our planet.

We can't live without the ocean

The ocean covers 71% of our planet. Or nearly three-quarters. And it is more than just a giant pond, that we visit on holiday. *We can't live without the ocean.* Let's see why.

▲ *One theory is that life began at ocean ridges, around* ***hydrothermal vents*** *like this one. It spurts out water full of chemicals, at up to 400 °C.*

1 It's where life began

Scientists tell us that life began in the ocean, around 3.8 billion years ago. But they are not yet sure exactly how.

- The first organisms were very simple. But they evolved over billions of years into a variety of ocean plants and animals.
- By around 500 million years ago, plants and animals had begun to leave the ocean, and adapt to life on land.
- Evolution continued on land, as well as in the ocean. (It still does.) And after 500 million more years, our human ancestors appeared.

We still show one link with the ocean. Our bodies need small amounts of ions found in ocean water. Sodium and potassium ions, for example.

2 It helps to regulate the atmosphere

The ocean supplies our atmosphere with oxygen. And helps to control the level of carbon dioxide. Follow the numbers on red to see how. Follow the others for extra information.

1 Up to half of the carbon dioxide we humans produce (for example by burning fossil fuels) is being absorbed by the ocean.

2 Tiny plants called **phytoplankton** use the carbon dioxide and water to make **glucose**, by **photosynthesis**. This needs sunlight.

They use the glucose, and nutrients (such as nitrogen and iron) from the water, to make the materials they need to grow.

3 Photosynthesis produces oxygen, which goes into the atmosphere. At least half of the oxygen in the atmosphere came from the ocean!

4 Meanwhile, little fish eat the phytoplankton. Big fish eat the little fish. So phytoplankton are the start of the food chains in the ocean.

5 Animal waste, and dead things, fall to the ocean floor. Some get buried in sediment, and in time turn into oil and gas. So carbon dioxide from the air ends up locked in these fossil fuels.

6 Bacteria and worms feed on some of the waste and dead remains. This releases carbon dioxide, and nutrients. These are carried up through the water by ocean currents. The phytoplankton recycle them.

3 It is the source of our fresh water

The water you drink, and wash in, started off in the ocean.

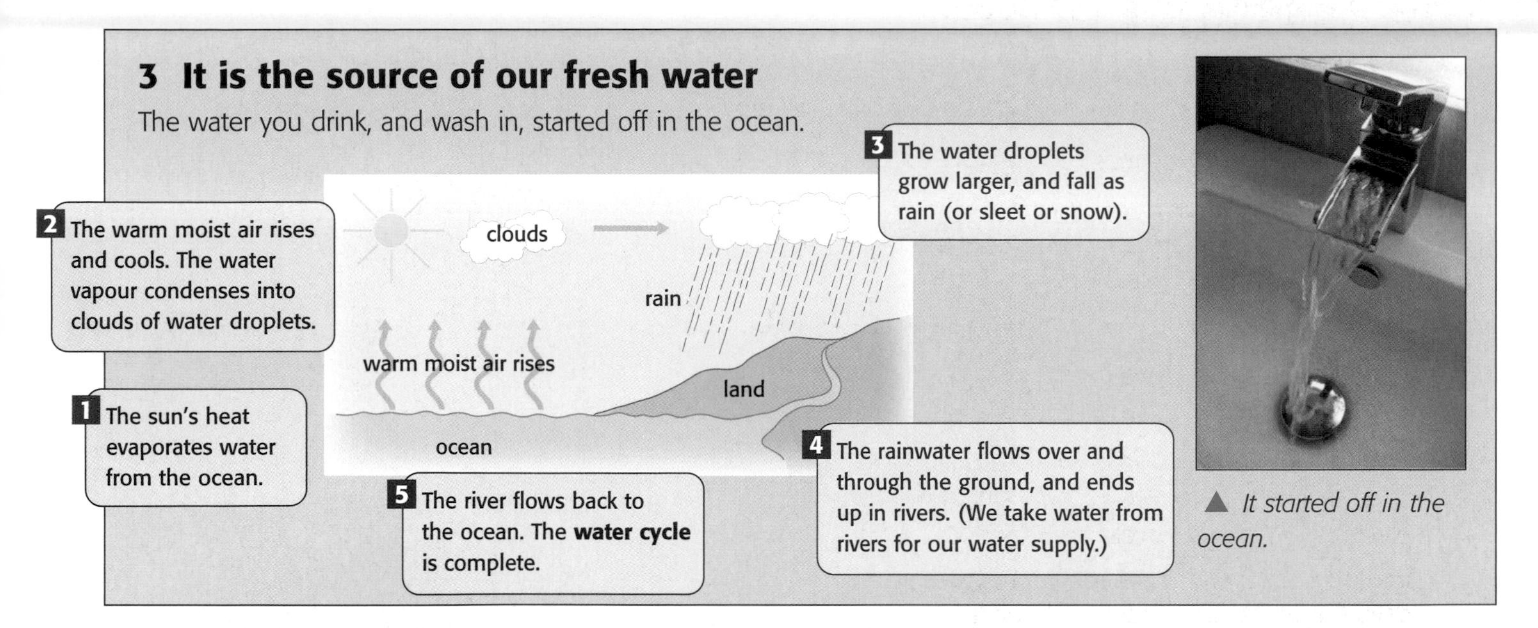

▲ *It started off in the ocean.*

4 It affects climates

The ocean affects climates around the world.

- First, it heats up, and cools down, more slowly than land does. So coastal places are cooler in summer, and milder in winter, than inland.
- It also helps to transfer heat from the Equator to colder regions. This affects the climates of coastal countries (like the UK).
- Two kinds of ocean current carry heat (and cold) around:
  - **surface currents**. These flow in the top 400 m of the ocean, dragged by winds. Your atlas may have a map of them.
  - **underwater currents**. These flow deeper and much more slowly.

The underwater currents form a system called the **global conveyor**. Look at this diagram.

It would take about 1000 years for a given 'parcel' of ocean water to complete the longer loop.

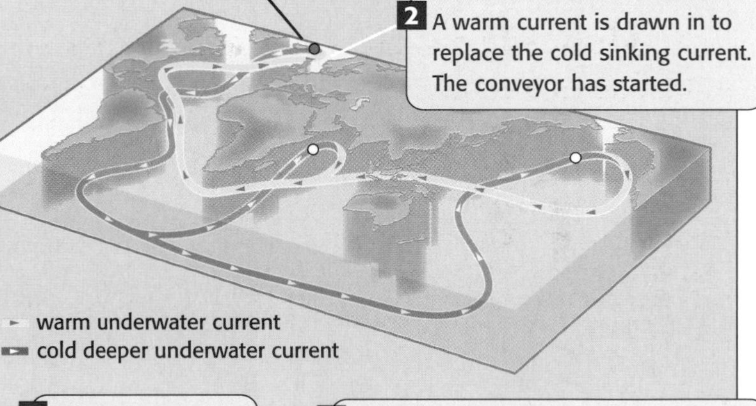

Your turn

1 Your connection with the ocean stretches back billions of years. Explain why.

2 The ocean absorbs carbon dioxide, which is then used by phytoplankton.
  - a What are phytoplankton?
  - b What do they use the carbon dioxide for?
  - c Phytoplankton can't grow in the deep ocean. Why not?
  - d In one way, phytoplankton are the most important species in the ocean. See if you can explain why.

3 Some of the carbon dioxide in the ocean gets locked up in fossil fuels. Will this be forever? Explain.

4 a Explain what the global conveyor is, in your own words.
  - b Starting at the Arctic, follow the conveyor through the longer loop. List the oceans it flows through, in order.

5 Now back to those wicked aliens, stealing our ocean. (See question 6 on page 118.) What impact will its loss have on us humans? Describe the impact as fully as you can. Use bullet points, or a consequence map.

7.3 How we use the ocean

Here you'll find out about ways we use the ocean, with the help of some clues.

An endless resource

We make use of the ocean in many different ways. You can probably think of some already.

Look at these photos for clues. Then try 'Your turn'.

STATEMENTS FOR QUESTIONS 5 AND 6

- **a** In many coastal countries, fish is the main source of protein.
- **b** But people began to protest against dumping.
- **c** Nations have fought each other at sea for thousands of years.
- **d** We are likely to extract other materials from the ocean in future.
- **e** We plan to use the power of the waves and tides to give electricity.
- **f** There are also oil and gas deposits on land, thanks to plate movements.
- **g** The London Protocol is now in force. It is a world-wide ban on dumping harmful things in the ocean.
- **h** Today, most coastal countries have a navy of some sort, for protection – even if it's just a few patrol boats.
- **i** At least half the magnesium we use (for making fireworks and alloys) is extracted from ocean water.
- **j** Ferries carry trucks, and cars, and people, on shorter sea journeys.
- **k** Each year, over 80 million tonnes of fish are taken from the ocean.
- **l** You can surf, swim, sail, snorkel, dive, and windsurf there.
- **m** Up to 40 years ago, we used the ocean as a legal dump. Factory waste, household rubbish, and all kinds of other things, were shipped out and chucked in.
- **n** Offshore windfarms make use of the strong ocean winds.
- **o** Most people love the ocean, and many of us head for the coast on holiday.
- **p** We extract oil and gas from under the ocean. It is easier where the water is not too deep – for example on the continental shelf.
- **q** Today, container ships and tankers carry goods and materials around the world.
- **r** In some countries they obtain salt by evaporating sea water. You can buy it as 'sea salt' in the shops.

Did you know?

A coastal country:
- 'owns' the ocean up to 22.2 km from its coast
- has the sole right to exploit the ocean floor up to at least 322 km from its coast.

CLUE BOX

- nuf
- lio nad ags
- isfhgni
- tarrspont
- sa a pumd
- rawfear
- banwerlee enegyr
- last nad rothe matersail

Your turn

The photos on page 122 show ways we use the ocean. The clue box above has matching clues – but jumbled!

1. First, make a *large* table like the one started below. (Leave room to write a lot in the third column.
2. Write all the letters **A** – **H** in the first column, to match the photos.
3. Now look at photo **A**.
   - **a** What use do you think it shows?
   - **b** See if you can find the matching term in the clue box. Unjumble it, and write it in the second column.

Photo	Things we use the ocean for	Notes about these uses
A		
B		
C		

4. Repeat **3** for all the other photos. (Easy ones first?)
5. Now look at the first use listed in your table. At least two of statements **a** – **r** above apply to it. Pick them out, and use them to help you write a logical note in your third column. You can add information of your own to the note, if you like.
6. Repeat **5** for all the other uses. (Make sure you have used all the statements **a** – **r**.) Well done!
7. Now, which of the eight uses do *you* think is:
   - **a** the most important? Explain why.
   - **b** the least important? Give your reasons.
8. Which use do you think has:
   a the most impact **b** the least impact
   on the ocean? Explain your choice each time.

Now for the bad news ...

Sadly, we are harming the ocean ecosystem. Find out more here.

Did you know?

- Harmful chemicals from industry are found in polar bears.
- They are carried to the Arctic by wind and ocean currents.

Our harmful impact on the ocean

You can think of the ocean as a vast ecosystem. We can't live in it – not like ecosystems on land. But we still manage to harm it. Below are ways in which we affect it.

1 Through burning fossil fuels

When we burn fossil fuels, carbon dioxide is emitted...

The level of **carbon dioxide** in the air is rising. That's because of all the fossil fuel we are burning – coal, oil, and gas.

So the ocean is absorbing more carbon dioxide. (There's always a balance between the amount in the air and the ocean.)

So the ocean water is slowly becoming more **acidic**. (Carbon dioxide is an acidic gas.)

This is making it more difficult for corals to form, and for shelled animals like crabs and clams to grow their shells.

It is helping to cause **global warming**: a rise in average air temperatures around the world.

So the ocean is warming up too.

So fish have already begun to move to cooler waters. Any that can't adapt will die.

And the warmer water is causing coral reefs to bleach, and die off.

So sea ice is melting in the Arctic and Antarctic.

This affects polar bears, seals, penguins, and other animals that live on or under the ice.

So land ice is melting around the Arctic and Antarctic.

The fresh water from it is flowing into the ocean, causing sea levels to rise.

The flow of fresh water at the Arctic could stop the global conveyor (page 121). That would have a drastic effect on the ocean ecosystem, and on world climates.

▲ *Warmer water and increasing acidity cause coral to bleach (lose its colour). It may slowly recover – or die.*

▲ *Polar bears hunt for seals on and at the edge of the ice. The more ice that melts, the harder it is to find food.*

2 Through overfishing

Around two million motorised fishing boats work the ocean. Most are quite small. But some are massive, with built-in fish factories, and can stay at sea for months.

The big growth in fishing has led to many problems.

- In some areas of the ocean, scarcely enough fish are left to breed. So fish stocks have collapsed.
- In many areas, stocks are close to collapse.
- So other species that feed on these fish are suffering too. Their food chains are breaking down.
- Some trawlers drag their nets along the ocean floor. This destroys the habitats of creatures that dwell there.
- Tons of *unwanted* species get caught in the fishing nets. They are thrown back into the ocean, dead.

Fishing for some species is now banned in some areas, to let stocks recover. But this does not really solve the problems.

▲ *The more the merrier?*

3 Through pollution

There is a world-wide ban on dumping harmful things in the ocean. Even so, it is still being polluted.

- Rivers carry stuff to the ocean. Like raw sewage, and fertiliser from farms, which cause a massive growth of phytoplankton. When they die, bacteria feed on them, using up the oxygen in the water. So fish may suffocate.
- Plastic waste gets carried in too. If sea animals eat it, it fills their stomachs. So they can't eat other things, and they starve. The plastic also gathers in certain areas, carried by currents. So you get miles of 'plastic soup'.
- Every so often, an oil tanker gets damaged, and oil spills into the ocean. Oil is toxic for ocean life.

▲ *You'll find this in the middle of the Pacific.*

Your turn

1 So far, the more carbon dioxide we have emitted, the more the ocean has absorbed.
  a This helps us. Why?
  b But it's starting to harm the ocean ecosystem. How?

2 Which do you think are responsible for overfishing?
  a just the coastal countries **b** most countries
  Explain your answer.

3 Which do you think are responsible for ocean pollution?
  a just the coastal countries **b** most countries
  Again, explain your answer.

4 Even people who live miles from the ocean, and rivers, are harming the ocean ecosystem. Explain why.

5 Look at those three ways in which we are harming the ocean ecosystem. Which do you think is the worst?
  a Put them in order, the most harmful first.
  b Explain why you chose that order.
  c Would a polar bear choose the same order as you?

6 *By harming the ocean ecosystem, we harm ourselves.* Do you agree with that statement? See how many reasons you can give, to back up your answer.

8 Our world in 2030

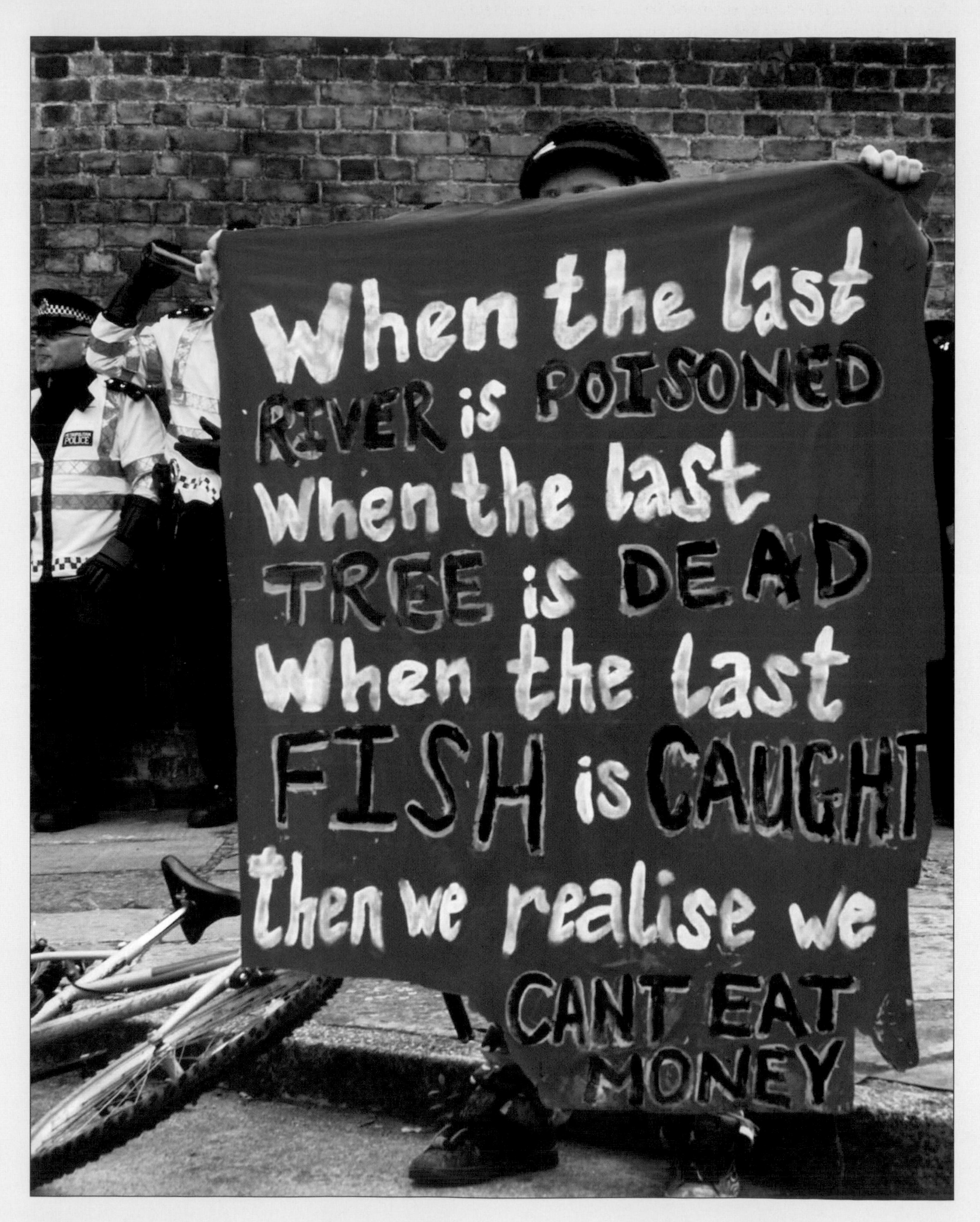

The big picture

This chapter is about how we need to start living sustainably.
These are the big ideas behind the chapter:

- We are making more demands on the Earth than it can support.
- In other words, we are living unsustainably.
- We can get an idea of our demands by calculating our ecological footprint.
- Our footprint is growing year after year, which means there are big problems ahead.
- We need to act quickly, to avoid these problems.

Your goals for this chapter

By the end of this chapter you should be able to answer these questions:

- Our ecological footprint takes into account the quantity of living resources (such as crops) that we consume. It also takes two other things into account. What are they?
- What can I say about the UK's footprint?
- The total human footprint is growing every year. Why? (Give three reasons.)
- At the same time, in many places, the Earth is becoming less able to support us. Why? (Give at least two reasons.)
- Because our footprint keeps on growing, we are heading for many problems. What kinds of problems? (Give at least three.)
- Suggest things governments, scientists, and individuals could do, to reduce our footprint. (Try for two sensible suggestions for each.)

And then ...

When you finish this chapter you can come back to this page and see if you have met your goals!

Did you know?

- The Earth's population is rising by over 70 million people a year.

What if ...

- ... we found another planet we could live on?

Did you know?

- We throw out over 30% of the food we grow and buy, here in the UK.
- The wasted food is worth over £8 billion a year.

What if ...

- ... there was not enough food for us all, here in the UK, by 2030?

Your chapter starter

Look at the photo on page 126.

Which country was the photo taken in?

What do you think is going on here?

Who do you think that message is aimed at?

See if you can put the same message in another way, in 12 words of your own.

Can everyone live like we do?

Is it possible for everyone to live like we in the UK do? Find out here.

It's an unfair world

It's an unfair world. 15% of the world's population live in abject poverty, with not even enough to eat. 24% do not have electricity.

Here, we are quite well off. The UK is in the top 20 wealthiest countries. So suppose you could wave a magic wand, and make the world more fair. Then could everyone live like we in the UK do? Let's see.

The way we live

Year after year, we rely on the Earth for the living resources listed here. Often they grow thousands of miles away, and we import the materials.

We depend on those resources for our food. And for much of our clothing. And for things like paper, and wooden furniture.

Year after year, we also rely on the Earth to absorb the waste we produce. In particular, the carbon dioxide gas from burning fossil fuels.

So could everyone live like we do? No! The Earth could not support it.

Our ecological footprint

Here in the UK, we make big demands on the Earth. We can get an idea of how big, by calculating our **ecological footprint** (or just **footprint**).

A country's footprint is the total area of the Earth's surface that is needed:

- to produce the crops and animals for its food supply
- to grow the trees for its timber and paper, and non-food crops (such as cotton) for fibre
- for its settlements, and roads, and other infrastructure
- to grow enough forest to soak up its carbon dioxide emissions, after the ocean has absorbed a share. (Trees and other plants take in CO_2.)

- So our footprint reflects the quantity of living resources we consume, *and* our living space, *and* our carbon dioxide emissions.
- The bigger the footprint, the bigger our demands on the Earth.
- The UK imports bananas from the Caribbean, and oranges from Brazil. The land these grow on is part of the UK's footprint.
- A footprint changes over time, for example as the population grows.

For 2005, the UK's footprint was 317.5 million **global hectares** (or **gha**).

That's **5.3 gha per person**. Or 53 000 sq m, or about 7 football pitches, for each one of us. (**Global** means it is land of world-average productivity.)

The trouble is …

The trouble is, the Earth was able to support a footprint of only **2.1 gha** per person in 2005, in a sustainable way.

- So if everyone lived like us in the UK, we'd have needed 2.5 Earths to support us. (5.3 ÷ 2.1 is roughly 2.5.)
- If they lived like people in the USA, we'd have needed 4.5 Earths.
- The *actual* average footprint for the world's population that year was **2.7 gha** per person – or *more than the Earth could support sustainably.*

So we used things up faster than the Earth could replace them, that year. For example, cut down trees at a faster rate than new trees could grow, and consumed fish stocks faster than new fish could breed.

We also emitted more carbon dioxide than the Earth could soak up – so it collected in the atmosphere.

Everyone should be able to live like us. It's only fair.

But our Earth can't support it,

For a start, global warming would rocket.

So what can we do?

▲ *The big dilemma…*

Mapping our ecological footprint

This map shows the footprint per person for different countries, for 2005. As you can see, it varied around the world. For many countries, it was much less than 2.1 gha per person. Lucky for us!

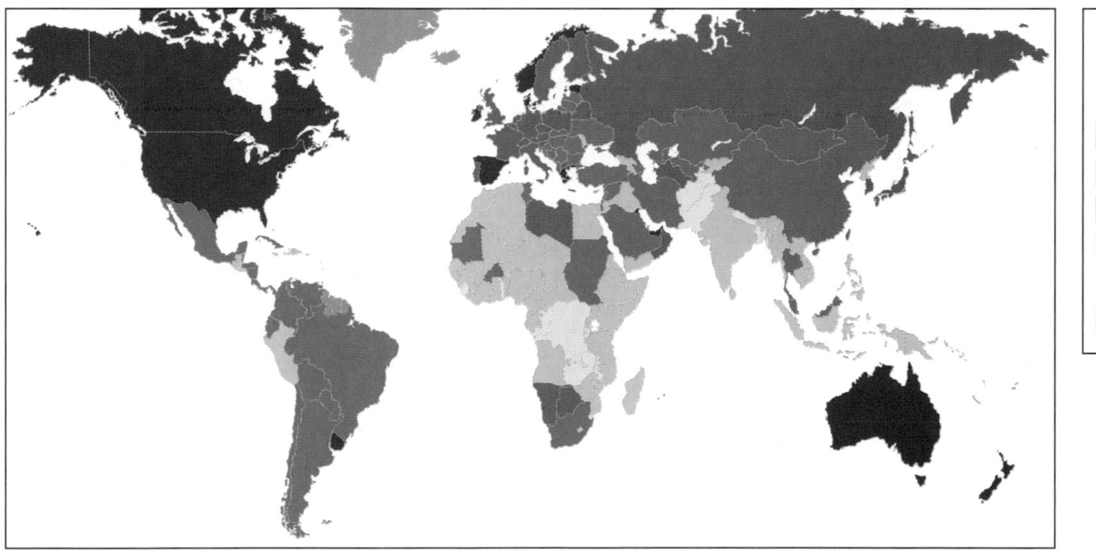

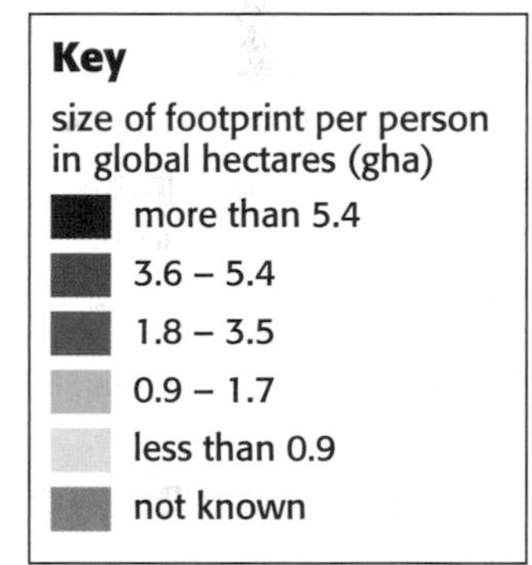

The main reason for the difference between countries is the difference in carbon dioxide emissions.

Your turn

1. How do you think each of these would affect the UK's total footprint? Each time, give reasons.
   - **a** Our population falls by half.
   - **b** We all start eating twice as much as before.
   - **c** We don't buy new clothes until the old ones are completely worn out.
   - **d** We leave all the street lights on all day.
   - **e** Farmers around the world begin to get huge crop yields, thanks to new types of seed.
2. See if you can think up two other changes that would affect our total footprint. Say what the effect would be.
3. Look at the footprint map.
   - **a** Name three countries in the group with the largest footprint per person. (Pages 140 – 141 may help.)
   - **b** What can you say about the footprint per person for: **i** Brazil? **ii** India? **iii** China?
   - **c** If everyone had a footprint like ours in the UK, it would be bad news. Explain why.
4. **a** Do you think there may be a link between footprint per person, and GDP per capita? Give reasons.
   - **b** Compare the map above with the one on page 16. Can you find evidence to back up your answer for **a**?

8.2

Trouble ahead!

This unit is about how the Earth is struggling to support us.

Our demands on a struggling Earth

As you saw, humans had a larger footprint in 2005 than the Earth could support sustainably.

But not only in 2005. Our footprint is still growing, year after year, and the Earth is struggling to cope.

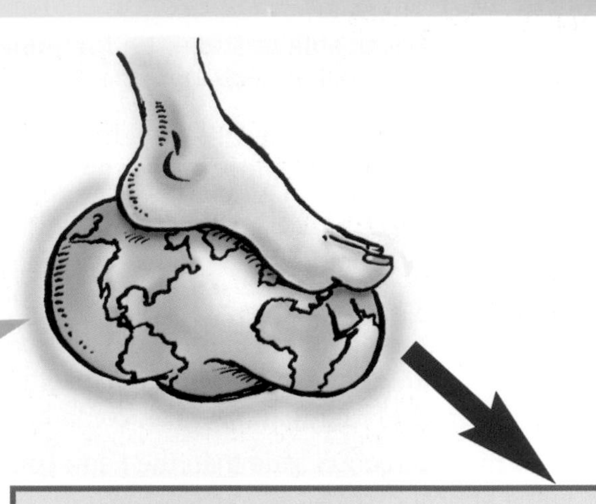

Our growing footprint

The total human footprint is growing. Because ...

the population is rising

- The Earth's population is rising by over 70 million people a year. Or by more than the population of the UK!
- They all need food, clothing, and space to live in.

we want more!

- Most of us like buying things, if we get the chance.
- Some LEDCs are developing fast. (For example China and India.) Their people are becoming better off. So they eat a wider variety of food, and buy more things.

carbon dioxide emissions are rising

- As people buy more cars, electrical goods, imported food, and other things, carbon dioxide emissions rise. (Think of all those factories and power stations burning fossil fuels, and transport burning petrol and diesel.)

A struggling Earth

Meanwhile, parts of the Earth are becoming less able to support us. Because ...

land is turning to desert

- In many places, farmland is turning to desert, thanks to overgrazing, intensive farming, and drought. This is a very big problem in parts of Africa and Asia.

fish stocks are collapsing

- In many parts of the ocean fish stocks are in great danger, due to overfishing.

rainforest is going

- Rainforest is still being destroyed, to clear land for crops and ranches. So the amount of farmland grows – but the exposed soil soon loses its goodness.
- And when the rainforest trees are burned, or left to rot away, they give off carbon dioxide.

climates are changing

- As the level of carbon dioxide in the air rises, the Earth gets warmer, and climates change.
- This will change patterns of farming and fishing around the world. Some countries may benefit, but many will suffer.

▲ *Our numbers keep on rising.*

Overgrazing can make land useless, as in this African village. ▶

Get ready for 2030!

What will the world be like in 2030? Beset by problems.

First, the world's population will have risen to over 8 billion. All will need feeding. There will probably be much less poverty. That's good news. But as people grow wealthier they like to eat richer and more varied diets. Demand for meat and dairy products will rise – so more farmland will be used for grazing instead of crops.

The larger population, and richer diets, means we will need about 50% more food than today. That's a massive challenge.

The demand for energy will have risen too. So, even if we make great strides with renewable energy, we will still depend on fossil fuels. Carbon dioxide emissions will still be too high. We will be struggling to keep global warming under control.

And there's another problem: water. Water is not a living resource, like crops or farm animals. But no living thing can survive without it. Today, farming uses about 70% of all fresh water. As the population rises, more water will be needed for growing food – and for drinking and cooking and washing. Demand for water will increase by about 30%. So expect conflicts over water use.

Climate change means some places will be much drier than now, and some wetter, with frequent flooding. This will have an impact on food production. If there's a shortage of food, or water, or both, in a place, what will people do? Move. Within their own country, or to another.

So, this is what we can expect by 2030:

- food and water shortages
- instability and conflict, because of the shortages
- climate change making those problems worse, in many places
- large numbers of people on the move, struggling for survival.

We must act now, to try to avoid these problems. It is not long until 2030.

Based largely on a speech by the government's chief scientific advisor, March 2009.

▲ *Sweets are okay for now. But in 2030, it may be food that's needed!*

Your turn

1 **A – J** below can be put into two groups.
Group 1: these increase our demands on the Earth.
Group 2: these lessen the Earth's ability to support us.
Which of **A – J** fit best in **Group 1**? Which fit best in **Group 2**? Find a good way to show your answer.

A Every couple chooses to have four children.
B A deadly virus wipes out the ocean's fish stocks.
C Southern Europe gets too dry for crops.
D All homes in Europe get air conditioning.
E The soil in much of the world loses its goodness.
F Rising sea levels drown all the low-lying coasts.
G We buy all the latest fashions, every season.
H In China, desertification increases rapidly.
I Almost all the Earth's forests have gone.
J In several countries, war puts an end to farming.

2 The total human footprint is much greater now than it was 200 years ago. See if you can explain why. Give as many reasons as you can.

3 Water is not a living resource, so it's not included in our footprint. But we can expect water shortages by 2030. Explain why.

4 The article above outlines problems the world will face in 2030. Do you think they could affect the UK? If yes, give some examples.

5 Do you agree with this comment? Decide on a response, and write it down.

> People should stop making these gloomy predictions. It just makes us feel bad.

Help us, somebody!

Who can help us to avoid the problems ahead? Explore that question here.

We are a clever species

We humans are clever. We have adapted to every climate, and ecosystem, on dry land. We have learned to exploit the oceans. We have walked on the moon.

So we ought to be able to find ways to reduce our footprint, and live on the Earth more sustainably. Yes? But who can help us?

Can our leaders help?

We expect the world's leaders to make the big decisions that will save us.

But they often find it hard to reach agreement. Look on the right.

Usually, their main aim is to help their economies grow – so that we all have more money to spend!

They think that's what we all want most. Could they be right?

Do we all need a new way of thinking?

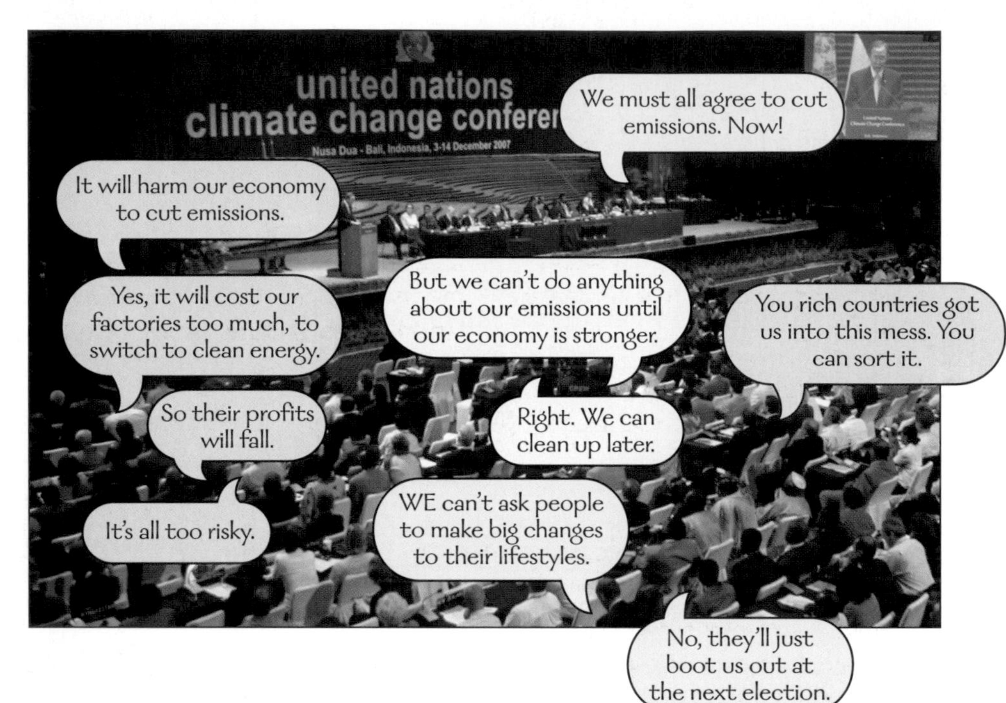

Can scientists help?

People blame scientists for some of the world's problems. But scientists are working hard to find solutions too. Here are just three examples:

They are helping farmers to restore useless soil. This can make a huge difference to food supplies. Planting trees, storing rainwater, and using fertiliser: all play a part.

Seeds are being modified to help crops cope with drought and other stresses. The result is **genetically modified** (or **GM**) crops. If proven safe, they may be very important.

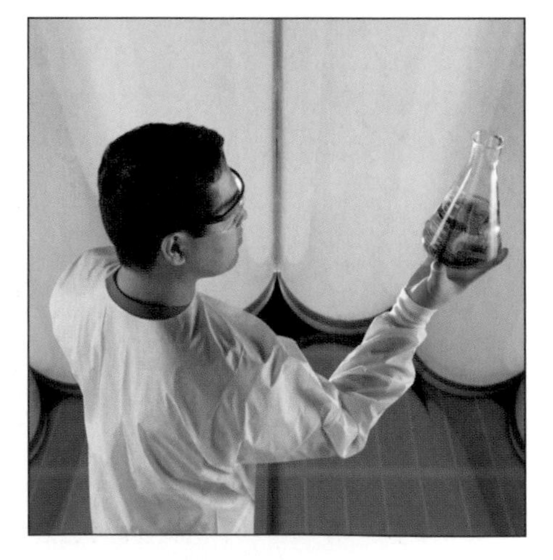

These hanging plastic sacks contain tiny plants called **algae**, in water. They grow fast. On crushing, they give the oil shown in the flask. It could become a key **biofuel**.

Can ordinary people help?

Many people are already reducing their own footprints on the planet.

For example, by cycling to work. Or using public transport instead of cars. And flying only if it's really necessary.

More and more of us recycle. It saves energy and materials – and means that less land is needed for landfill sites to bury our rubbish in.

Some people organise or take part in protests, to remind us all of the issues. It is also a way to put pressure on governments.

Mostly, people are trying to reduce their carbon dioxide emissions. That's fine, because these account for a large part of our footprint.

Your turn

Perhaps **A – O** on the right could help us avoid those problems predicted for 2030. (See page 131.)

1 From **A – O**, pick out all the actions that you think:
- **a** would depend mainly on leaders and governments
- **b** would depend mainly on scientists
- **c** are down to ordinary people
- **d** would probably not help at all

To answer, you can just give their letters (shown in pink).

2 From your list for **1a**, pick out an action that you think:
- **a** would cause the most resentment
- **b** would cause the least resentment
- **c** is the most essential for tackling world problems
- **d** could reduce our footprint, but cause more poverty in developing countries

3 Now look again at **A – O**. Pick out the things that you feel quite sure will *never* happen.

4 **a** So far, who do you think has *the most power* to tackle the problems predicted for 2030?
governments scientists ordinary people
b Put the three groups in **a** in order of power, the most powerful first. Explain the order you chose.

5 *'It's more important to reduce our footprint than get rich.'* How much do you agree with that statement? Choose a number from 1 to 5.
(1 = strongly disagree, 3 = neutral, 5 = strongly agree)

A Grow vegetables in the garden.

B Get all countries to work on the problems together.

C Do not waste food.

D Put a big tax on flights.

E Find a way to grow crops in the ocean.

F Stop all trade with other countries.

G Pass a law that women can have no more than one child.

H Put solar heating panels on the roof.

I Give away the things you don't use.

J Get people to work just four days a week.

K Develop space stations for humans to live in, on other planets.

L Do without a car.

M Allow each person 1 flight every 5 years.

N Create food pills in the lab. No more farms needed!

O Bear footprint in mind, when shopping.

6 Imagine you live in Burkina Faso. (See the lower left photo on page 96.) Will your answer for **5** be the same? Give your reasons.

7 Look back at the message on the banner, on page 126. Then think again about your answers for **5** and **6**.
- **a** Do you want to change either answer now?
- **b** Might your answers be different in 2030?

8.4 You: part of the solution?

This unit explores what you could do, to help the world tread lightly.

You in 2030

In 2030 you'll be an adult. You may even have children of your own. So what kind of world do you want, for you and your children? That is a very serious question.

The challenge

Look at this graph. In 2005, we really needed 1.3 Earths to support us. If we carry on like this, along the red line, we'll need almost 2 by 2030.

But we have only 1 Earth. So we'll be in big trouble. We'll have all the problems you read about on page 131.

Key for graph

1961 – 2005
- actual footprint

2005 onwards
- carry on as usual
- reduce footprint fast

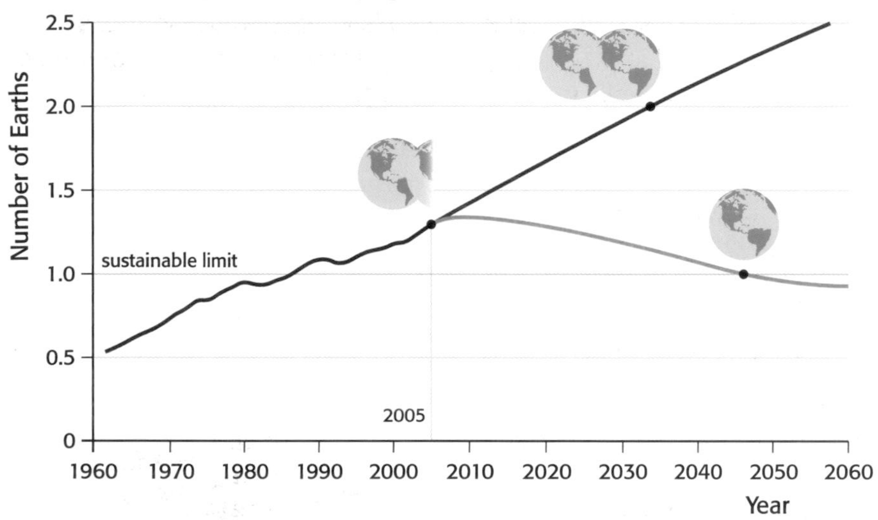

Or we could follow the yellow line. We could start reducing our footprint now. And fast. We could be living sustainably on our Earth before 2050. Which path would you choose?

▲ *Their future depends on us.*

Can you make a difference?

You have every right to be on this Earth, and to have a say in its future. But you may feel that you can't make a difference.

Think again. Suppose 100 people like you take action. Then 1000 … 10 000 … 100 000 … 1 million. Change is on the way.

And suppose the 1 million grew to 5 million… 10 million… 1 billion… 2 billion… 3 billion … 6 billion… Job done. Easy!

It might be only a small action, that reduces our footprint by a fraction. But if enough people join in, the effect can be enormous.

Can geography help?

It can sometimes be hard to decide on the best way to reduce our footprint. For example, should we go for nuclear power? No CO_2 – but it brings other risks.

That's where geography comes in. It helps you understand how the real world works. It helps you think about the issues, and look at the choices, and make up your mind.

In fact, geography is all about our impact on the Earth, and the Earth's impact on us. So it is a really important subject – and now more than ever.

Can it be fun?

You can help to reduce our global footprint. That way, you are part of the solution. (Otherwise, you are part of the problem!)

Can reducing our footprint be fun? Yes. Think about this:

- Our future is at stake – and so is the future of many other species.
- Ignoring the problems ahead will just make them worse.
- We humans have made some big mistakes. But overall, we are clever. We will find solutions we can't even imagine yet.
- Solving problems is a challenge. And challenges are exciting! Finding ways to live sustainably is the biggest challenge of all.
- Even small actions count. And who knows? You could go on to big important actions – inspired by what you learn in geography.

Your turn

1 Look at the graph on page 134.
  a How many Earths did we need to support our demands: i in 2005? ii in 1970?
  b In about which year did our demands first *exceed* the Earth's ability to support them?
  c The red line is sloping upwards. See if you can give three reasons to explain the predicted rise.

2 a Look at the yellow line on the graph. If we follow it, the Earth will be able to support us sustainably again. In about which year might this happen?
  b Give two things you could do, to help us get onto that yellow line.
  One should be something you'd find very easy.
  The other should be something you'd find really hard.

3 Our growing footprint affects other species too. For example the number of giant pandas fell because humans took over much of their habitat. The pandas were driven out – or hunted for their fur.
Imagine you are the mother panda, on page 134. Make a speech to persuade humans to get on that yellow graph line fast.

4 One student came up with this slogan for studying geography: '*Geography helps you go green*.'
  a What does it mean?
  b Do you think it's true?
  c See if you can come up with three reasons why everyone should study geography.
  d Now see if you can come up with a better slogan.

Ordnance Survey symbols

ROADS AND PATHS

- M I or A 6(M) — Motorway
- A 35 — Dual carriageway
- A31(T) or A35 — Trunk or main road
- B 3074 — Secondary road
- Narrow road with passing places
- Road under construction
- Road generally more than 4 m wide
- Road generally less than 4 m wide
- Other road, drive or track, fenced and unfenced
- Gradient: steeper than 1 in 5; 1 in 7 to 1 in 5
- Ferry — Ferry; Ferry P – passenger only
- Path

PUBLIC RIGHTS OF WAY

(Not applicable to Scotland)

1:25 000   1:50 000

- Footpath
- Road used as a public footpath
- Bridleway
- Byway open to all traffic

RAILWAYS

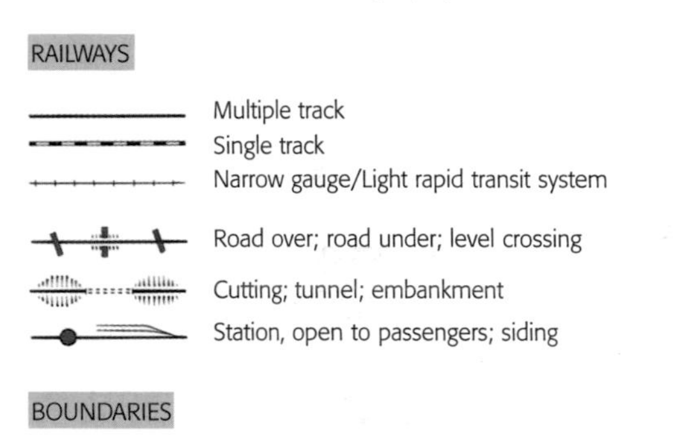

- Multiple track
- Single track
- Narrow gauge/Light rapid transit system
- Road over; road under; level crossing
- Cutting; tunnel; embankment
- Station, open to passengers; siding

BOUNDARIES

- National
- District
- County, Unitary Authority, Metropolitan District or London Borough
- National Park

HEIGHTS/ROCK FEATURES

- 50 — Contour lines
- ·144 — Spot height to the nearest metre above sea level

outcrop, cliff, scree, 650, 600

ABBREVIATIONS

P	Post office	PC	Public convenience (rural areas)
PH	Public house	TH	Town Hall, Guildhall or equivalent
MS	Milestone	Sch	School
MP	Milepost	Coll	College
CH	Clubhouse	Mus	Museum
CG	Coastguard	Cemy	Cemetery
Fm	Farm		

ANTIQUITIES

- VILLA — Roman
- Castle — Non-Roman
- Battlefield (with date)
- Tumulus

LAND FEATURES

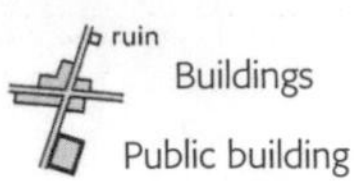

- Buildings
- Public building
- Bus or coach station
- Place of Worship — with tower; with spire, minaret or dome; without such additions
- Chimney or tower
- Glass structure
- Heliport
- Triangulation pillar
- Mast
- Wind pump / wind generator
- Windmill
- Graticule intersection
- Cutting, embankment
- Quarry
- Spoil heap, refuse tip or dump
- Coniferous wood
- Non-coniferous wood
- Mixed wood
- Orchard
- Park or ornamental ground
- Forestry Commission access land
- National Trust – always open
- National Trust, limited access, observe local signs
- National Trust for Scotland

WATER FEATURES

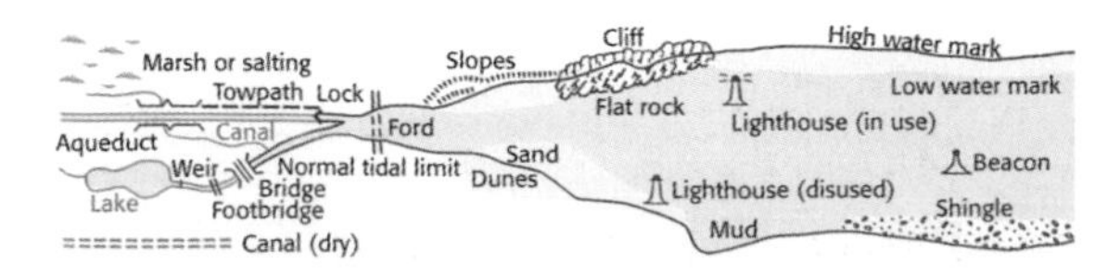

TOURIST INFORMATION

- Parking
- Visitor centre
- Information centre
- Telephone
- Camp site/ Caravan site
- Golf course or links
- Leisure/sports centre
- PC — Public convenience
- Picnic site
- Pub/s
- Museum
- English Heritage
- Steam railway
- Garden
- Nature reserve
- Water activities
- Fishing
- Walks/trails
- Other tourist feature
- Moorings (free)
- Electric boat charging point

Map of the British Isles

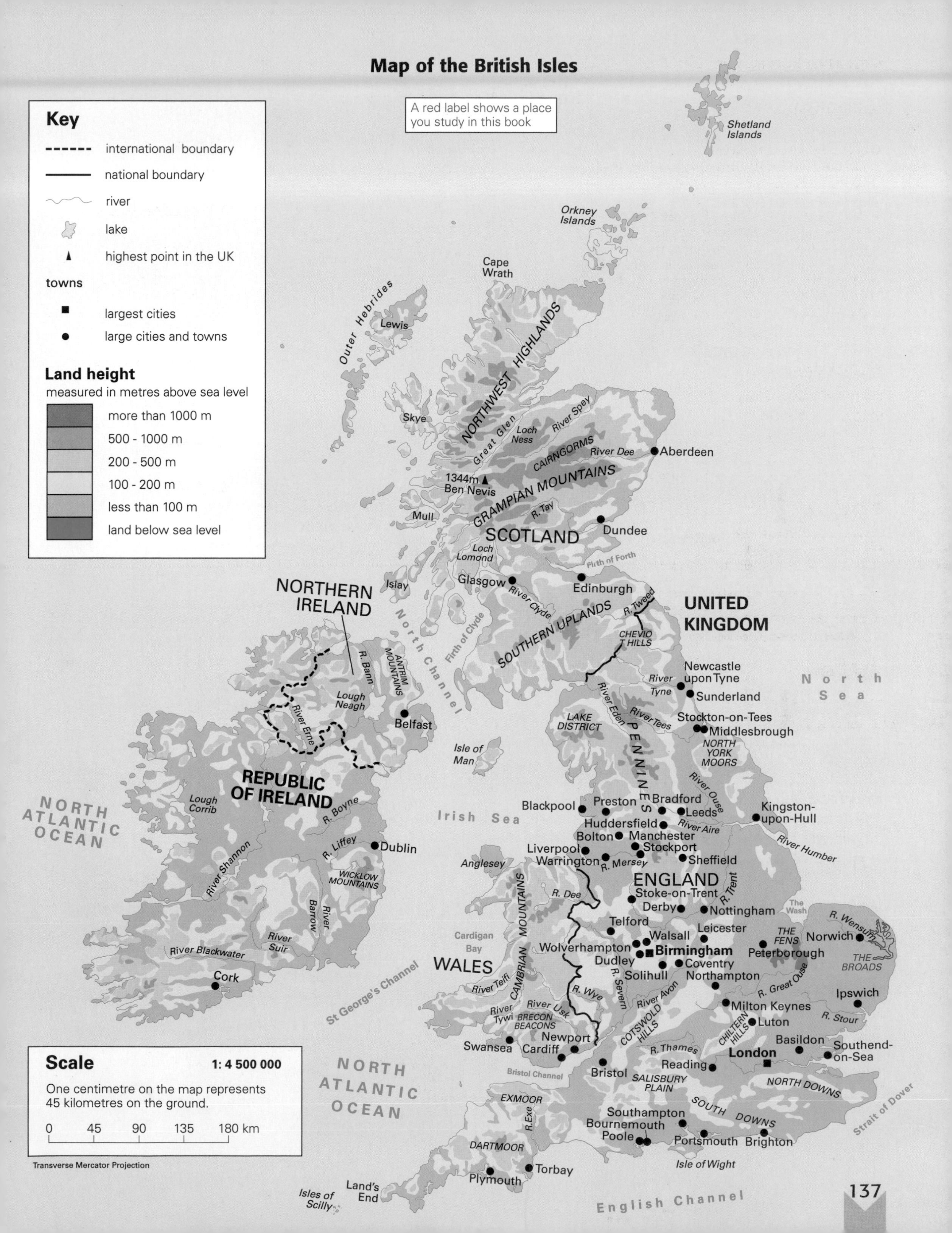

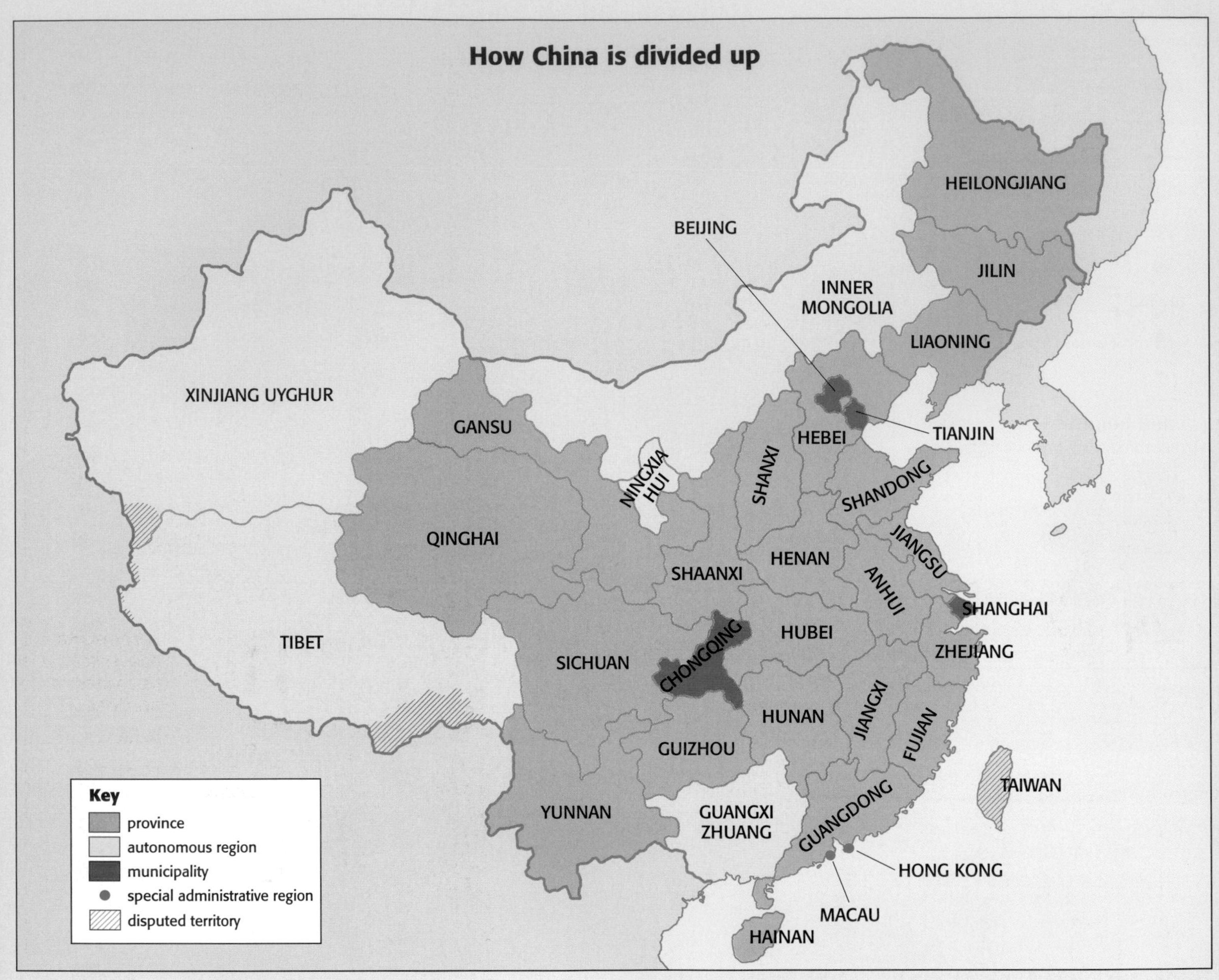

Hints on pronunciation

These will help you pronounce the names on the map.

Chinese	pronounce it
ch	like ch in teach
ei	as in eight
i	as in machine
iao	like yow in yowl
in	like in
j	like j in jeep
q	like ch in chair
x	like sh in she
ng	as in sing
uyghur	like wi-ger
zh	like j in just

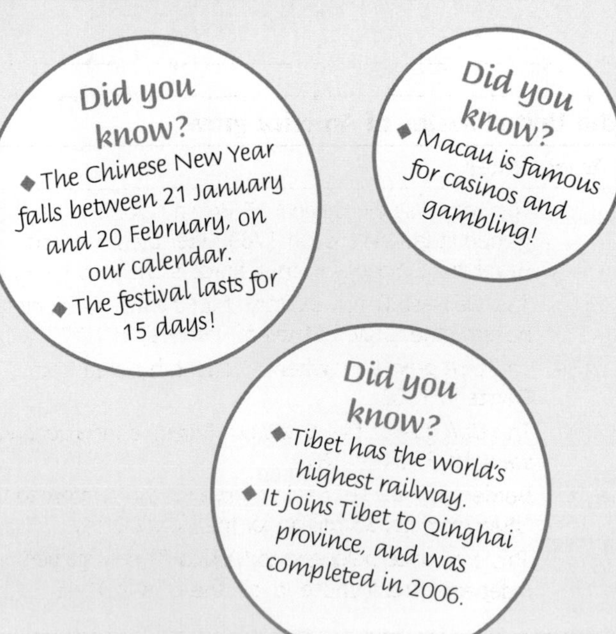

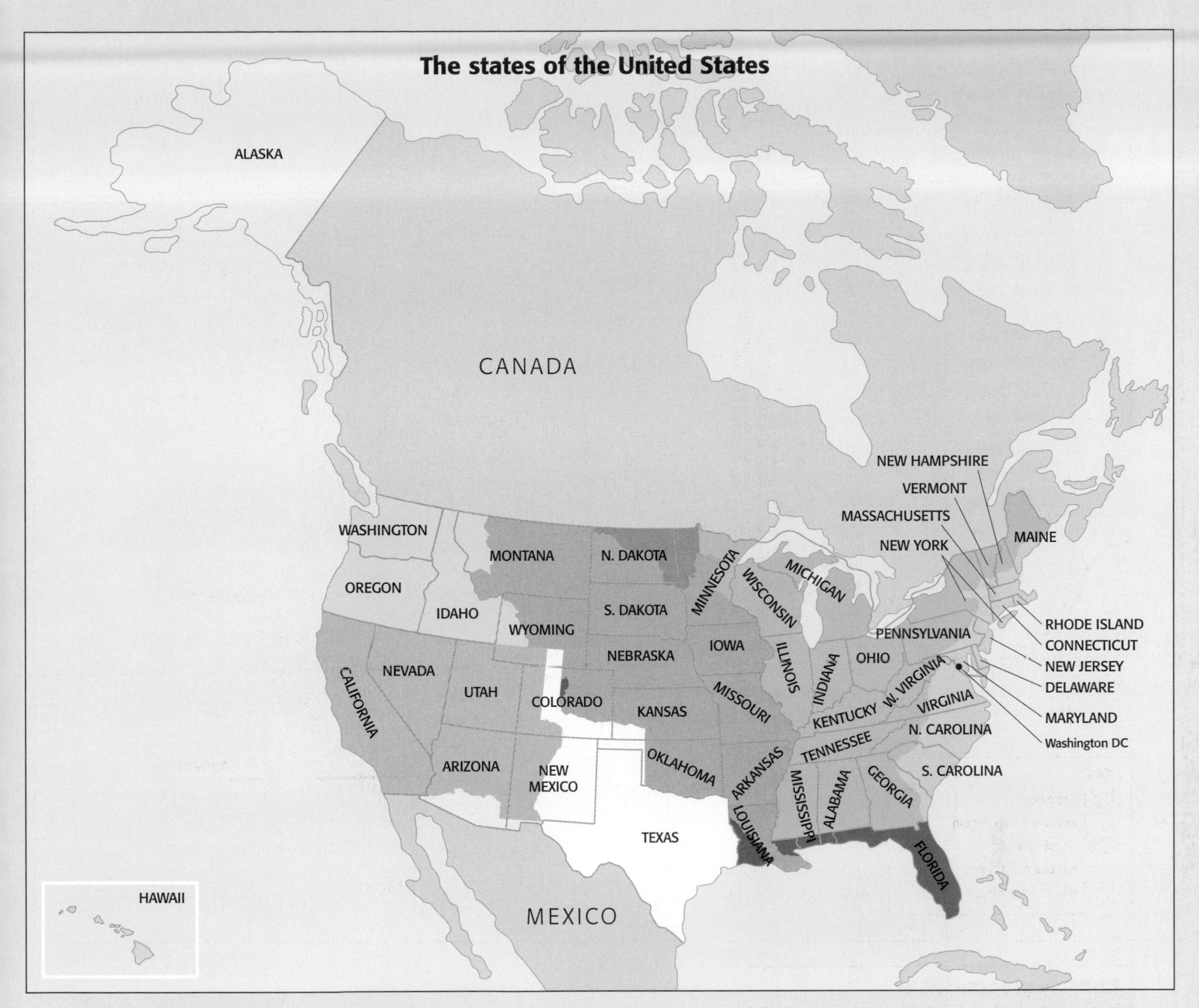

How the United States of America grew

Date	Area	Notes
1783		This area was the original 13 British colonies; they gained independence in 1783, after the American Revolution. (Several of their borders changed later.)
1783		This was also British territory. With the 13 ex-colonies, it became the 'United States of America', in 1783.
1803		The USA grew a lot when it bought this area from France in 1803.
1818		The USA gained this area from Britain, after a second war with Britain in 1812.
1819		Border disputes led Spain to give up this territory to the USA; Spain got $5 million for the Florida area.
1845		This area once belonged to Mexico. Then it gained independence. It chose to join the USA in 1845.

Date	Area	Notes
1846		After years of dispute, the British signed a treaty that gave this area over to the USA, in 1846.
1848		After a war with Mexico, the USA took this area over from Mexico in 1848. Mexico was paid $15 million in compensation for the war.
1853		The USA bought this area from Mexico for $10 million, in 1853.
1867		The USA bought Alaska from Russia in 1867, for $7.2 million.
1900		After years of conflict, and invasion by US forces, Hawaii was officially taken over by the USA in 1900.

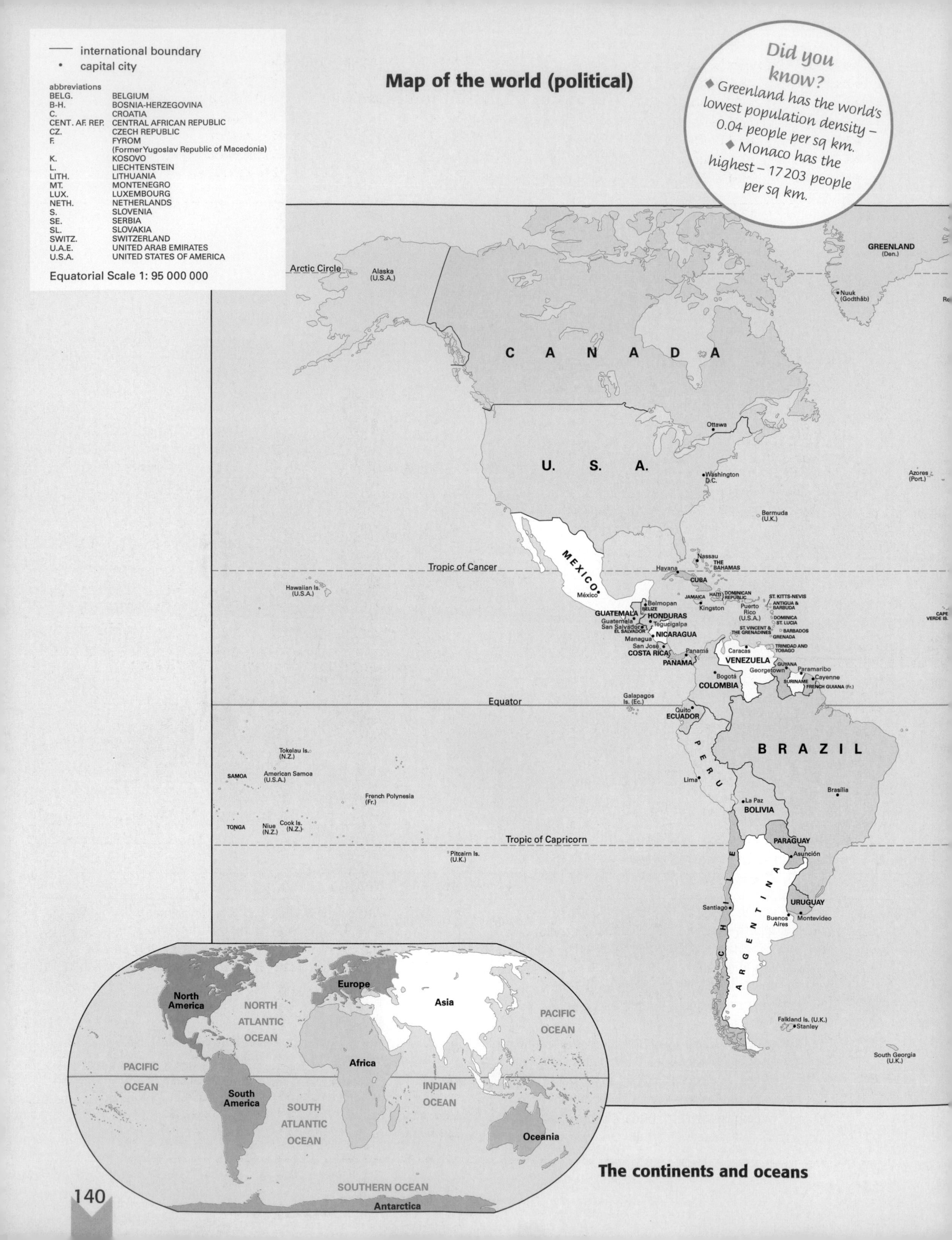

international boundary
capital city
abbreviations
BELG. BELGIUM
B-H. BOSNIA-HERZEGOVINA
C. CROATIA
CENT. AF. REP. CENTRAL AFRICAN REPUBLIC
CZ. CZECH REPUBLIC
F. FYROM
(Former Yugoslav Republic of Macedonia)
K. KOSOVO
L. LIECHTENSTEIN
LITH. LITHUANIA
MT. MONTENEGRO
LUX. LUXEMBOURG
NETH. NETHERLANDS
S. SLOVENIA
SE. SERBIA
SL. SLOVAKIA
SWITZ. SWITZERLAND
U.A.E. UNITED ARAB EMIRATES
U.S.A. UNITED STATES OF AMERICA
Equatorial Scale 1: 95 000 000
Map of the world (political)
Did you know?
Greenland has the world's lowest population density – 0.04 people per sq km.
Monaco has the highest – 17 203 people per sq km.
Arctic Circle
Alaska (U.S.A.)
GREENLAND (Den.)
Nuuk (Godthåb)
CANADA
Ottawa
U. S. A.
Washington D.C.
Azores (Port.)
Bermuda (U.K.)
MEXICO
México
Nassau
THE BAHAMAS
Tropic of Cancer
Havana
CUBA
Hawaiian Is. (U.S.A.)
JAMAICA
HAITI
DOMINICAN REPUBLIC
Kingston
Puerto Rico (U.S.A.)
ST. KITTS-NEVIS
ANTIGUA & BARBUDA
DOMINICA
ST. LUCIA
BARBADOS
GRENADA
ST. VINCENT & THE GRENADINES
TRINIDAD AND TOBAGO
CAPE VERDE IS.
Belmopan
BELIZE
GUATEMALA
Guatemala
HONDURAS
Tegucigalpa
San Salvador
EL SALVADOR
NICARAGUA
Managua
San José
COSTA RICA
Panamá
PANAMA
Caracas
VENEZUELA
GUYANA
Georgetown
Paramaribo
SURINAME
Cayenne
FRENCH GUIANA (Fr.)
Bogotá
COLOMBIA
Galapagos Is. (Ec.)
Equator
Quito
ECUADOR
BRAZIL
PERU
Lima
Tokelau Is. (N.Z.)
SAMOA
American Samoa (U.S.A.)
French Polynesia (Fr.)
TONGA
Niue (N.Z.)
Cook Is. (N.Z.)
La Paz
BOLIVIA
Brasília
Tropic of Capricorn
PARAGUAY
Asunción
Pitcairn Is. (U.K.)
CHILE
ARGENTINA
Santiago
URUGUAY
Buenos Aires
Montevideo
Falkland Is. (U.K.)
Stanley
South Georgia (U.K.)
North America
Europe
Asia
NORTH ATLANTIC OCEAN
PACIFIC OCEAN
Africa
PACIFIC OCEAN
South America
INDIAN OCEAN
SOUTH ATLANTIC OCEAN
Oceania
SOUTHERN OCEAN
Antarctica
The continents and oceans

Population of the world's continents

- Asia 4.05 billion
- Africa 0.97 billion
- Europe 0.74 billion
- N America 0.53 billion
- S America 0.38 billion
- Oceania 0.04 billion

The world's top five languages

	(native speakers)
◆ Chinese (Mandarin)	over 1 billion
◆ Hindi	498 million
◆ Spanish	391 million
◆ English	512 million
◆ Arabic	245 million

The G8 leading industrial nations

Canada
France
Germany
Italy
Japan
Russia
UK
USA

The G20 major economies

Argentina
Australia
Brazil
Canada
China
France
Germany
India
Indonesia
Italy
Japan
Mexico
Russia
Saudi Arabia
South Africa
South Korea
Turkey
UK
USA
EU

The G20 developing nations (more than 20!)

Argentina
Bolivia
Brazil
Chile
China
Cuba
Ecuador
Egypt
Guatemala
India
Indonesia
Mexico
Nigeria
Pakistan
Paraguay
Peru
Philippines
South Africa
Tanzania
Thailand
Uruguay
Venezuela
Zimbabwe

Glossary

A

acid rain – rain with acidic gases dissolved in it; it can kill fish and plants

adult literacy rate – the % of people aged 15 and over who can read and write a simple sentence

agribusiness – where farms are part of a much larger business, that may include factories making fertiliser or food

aid – help given to poorer countries

algae – tiny plants that grow in water

American Dream – the belief that you can succeed in America, no matter what your background is

aquifer – an area of rock below the ground, that holds water like a sponge

B

bacteria – tiny organisms, each just one cell; some are harmless, some cause disease

bilharzia – a disease caused by worms found in river water; it can damage your brain

biofuel – a fuel obtained from plants

C

call centre – where people work all day long on the phone (for example in telephone banking)

catering – providing food and drink

cholera – a disease caused by bacteria in dirty drinking water (you vomit, and have diarrhoea, and can die from dehydration)

colony – a country taken over and ruled by another country

commodity exchange – a trading centre where commodities like coffee and sugar are bought and sold on the world market

commune – where people share land and work on it together

communism – a political system based on the belief that people should not own anything; the state should own everything and give people what they need

continental shelf – where the ocean floor slopes gently away from the coast, before it plunges to the deep ocean

coral reef – a structure in the ocean, formed by limestone secreted by animals; reefs are home to many ocean species

D

dam – a structure built across a river, to control the flow of water; it usually has turbines, to give electricity

debt relief – cancelling a country's debts

deforestation – when forests are cut down

delta – flat area of deposited material at the mouth of a river, where it enters the sea

demand – the amount of something that people are willing to buy

desert – an area that gets under 250 mm of rain a year; a desert can be hot or cold

desertification – when soil in a dry region gets dried out, and useless

development – a process of change to improve people's lives

development indicators – data used to show how developed a country is

domestic tourist – being a tourist in the country you live in

dynasty – where the same family rules for generations

E

eco-friendly – does little or no harm to the environment

ecological footprint – a measure of how much of the Earth we use to support us

economic – to do with the economy, money, and earning a living

ecosystem – a unit made up of living things and their non-living environment (soil, warmth, water, and so on)

ecotourism – tourism based on wildlife and local culture; it aims to benefit the local people, and protect the environment

emissions – waste gases given out

environmental – to do with the environment (air, soil, water, wildlife, and so on)

EU (European Union) – the 'club' of European countries that have signed agreements on trade and other issues

F

fair trade – where the producer of the goods gets a fair share of the profits

Fairtrade Foundation – the body that allows companies to use the Fairtrade logo

free trade – when countries trade freely with each other, with no restrictions

fertilisers – are put on soil to help crops grow

G

GDP (gross domestic product) – the total value of all the goods and services produced in a country in a year

GDP per capita – the GDP divided by the population: it gives you an idea of how wealthy the people are, on average

genetically modified (GM) – its genes have been altered in the lab

global conveyor – a system of underwater warm and cold currents, in the ocean

globalisation – how companies, trade, ideas and lifestyles are spreading more easily around the world

global warming – temperatures around the world are rising

GNI (gross national income) – the total earned in a year in a country (including money in from other countries), minus what it paid out to other countries

groundwater – rainwater that has soaked into the ground and is held in rocks

H

HDI (human development index) – a 'score' between 0 and 1, to indicate how developed a country is; it combines data on GDP per capita, adult literacy, life expectancy, and enrolment in education; the higher the number, the better

heavily indebted countries – poor countries with large loans they can't repay

hemisphere – half of the globe; the northern hemisphere is north of the equator

Hispanic – a general term for people from Mexico and Central and South America

hydroelectricity – electricity generated when running water spins a turbine

I

IMF (International Monetary Fund) – a fund set up by governments to make loans to countries, especially for trade

inbound tourist – a tourist coming in from another country

in decline – coming to an end, dying away

illegal immigrant – a person who enters a country without official permission

immigrant – a person who moves into a country from another country

immigrant – a person who moves into a country from another country

indigenous people – the very first people to settle in a country, or region

infant mortality – the number of babies out of every 1000 born alive, who die before their first birthday

infrastructure – roads, railways, water and electricity supplies, and other basic systems a country needs, to function

interest – you pay this if you take out a loan

interest rate – the charge for taking a loan, given as a % of the loan (eg 5% per year)

international tourist – a tourist from another country

irrigation – watering crops

L

LEDC – less economically developed country (one of the poorer countries)

life expectancy – how many years a new baby can expect to live, on average

M

machete – a broad heavy knife used for harvesting crops, and other tasks

magma – melted rock below the Earth's surface

manufacturing – making things in factories

mass tourism – when a place becomes a destination for large numbers of tourists

mechanised – uses machines to do most of the work

MEDC – more economically developed country (one of the richer countries)

migrant workers – move from one area or country to another, looking for work

Millennium Development Goals – goals, agreed by world leaders, to reduce poverty in the world by the year 2015

millet – a type of cereal crop

N

National Park – a large area protected by law for the benefit of everybody

natural resources – resources that occur naturally, like oil wells and fertile soil

NGO (non-governmental organisation) – an organisation such as Oxfam, that is independent of the government

NIC (newly industrialised country) – has recently set up a lot of industry

O

ocean current – a current of warm or cold water flowing within the ocean

ocean ridge – a mountain ridge on the ocean floor, formed by rising magma

ocean trench – a deep chasm in the ocean floor, where one plate dives under another

one-child policy – women in China are encouraged to have only one child

P

package holiday – where you pay in advance for travel and accommodation

patent – when you register an invention to protect it, and prevent people copying it

permafrost – the soil that is permanently frozen, under the surface in the tundra

pesticides – chemicals sprayed on crops, to kill insects that would eat them

photosynthesis – the process in which plants make their food from carbon dioxide and water, in sunlight

phytoplankton – tiny ocean plants

political – to do with how a country or area is governed, or run

poor south – a term sometimes used for poorer countries (since many are in the southern hemisphere)

population density – the average number of people per square kilometre

population distribution – how the population is spread around the country

PPP (purchasing power parity) – when a figure (such as GDP) is adjusted to take into account that a sum of money buys more in some countries than others

precipitation – water falling from the sky in any form: as rain, hail, sleet, or snow

primary sector (of the economy) – people are employed in collecting things from the earth (farming, fishing, mining)

processing – converting a material from one form to another (for example cotton to denim, or milk to cheese)

profit – what's left when you subtract the cost of a thing from the price you sold it for

pull factors – factors that attract people to a place (for example, better wages there)

push factors – factors that push people out of a place (for example, no work there)

PV cell – cell that converts sunlight straight into electricity; it provides solar power

R

raw material – material to be processed; eg cotton fibre to be woven into cotton

recession – when economic activity in a country slows down; companies may close, and people lose their jobs

relief – the shape of the land (how high or low it is)

reservation /reserve – land set aside for a special purpose, for example for Native Americans, or rainforest tribes, to live on

revenue – money you take in from selling goods and services

rich north – a term sometimes used for the richer countries (since most are in the northern hemisphere)

rural – to do with the countryside

S

secondary sector (of the economy) – people work in manufacturing

service sector – see *tertiary sector*

smog – a haze that forms when pollutants in the air react, triggered by sunlight

social – to do with the way people live

solar power – uses energy obtained directly from sunlight; see *PV cell*

subsidy – grant (eg for growing a crop)

superpower – a big country with a strong economy, strong military, and great political and cultural influence

supply – the amount of something for sale

sustainable – can be continued into the future without harm

sustainable development – development that brings social, economic and environmental benefits

sustainable tourism – where local people have a say, and share the benefits, and the environment is protected

sweatshop – a factory where people work for long hours, for low pay

T

tariff – a tax a government places on imports (and sometimes, on exports)

tertiary sector (of the economy) – people provide services (like medical care, education, entertainment)

Third World – a name sometimes used for the world's poorer countries

Third World debt – the money owed by the poorer countries to the richer ones

TNC (transnational corporation) – company with branches in many countries

tornado – a violent spinning windstorm

tourism – everything to do with tourists, including the activities they take part in and the services that support them

tourist – a person who stays for more than a day in a place that is not his or her usual environment, for any purpose

tundra – the ecosystem around the Arctic, where the climate is very harsh, and the soil below the surface always frozen

typhoid – a disease caught from drinking dirty water; you get fever and may die

U

under-5 mortality rate – the % of babies born alive who die before they reach five

undernourished – when you don't get enough food to live a normal healthy life

United Nations (UN) – it aims to promote world peace, prosperity and justice; most countries of the world belong to it

United Nations Security Council – a panel within the UN, that deals with conflicts between, and within, countries

urbanisation – when rural areas become built up, as towns and cities spread

urban sprawl – when a city spreads in an uncontrolled way

W

water table – the upper surface of the groundwater held in rocks

wildfire – an outdoor fire that spreads very rapidly, for example in a forest

World Bank – a joint bank owned by governments of over 180 countries, set up to provide loans for development

WTO (World Trade Organisation) – set up to make trade between countries easier; over 150 countries belong to it

Index